Community-Owned Transport

City and state governments around the world are struggling to achieve environmentally sustainable transport. Economic, technological, city and transport planning and human behaviour solutions are often hampered by ineffective implementation. So attention is now turning to institutional, governmental and political barriers. Approaches to these implementation problems assume that transport ownership can only be public (owned by state entities) or private (corporate or personal). Another option – largely unexplored to date – is communal ownership of transport.

Community-Owned Transport proposes and develops the notion that communal ownership has a historical basis and provides unique opportunities for providing personal mobility. It looks at the historical roots of modern urban transport's failings as those of technological change and the associated governing of transport systems, particularly the role of public sector institutions. Community ownership is explored through the new 'sharing economy' developments – car sharing, ridesharing and bicycle share schemes – and older social innovations in ecovillages and communal living. Models and practices of community ownership of transport are provided and this study also discusses how community ownership might contribute to sustainable transport.

Drawing widely on different disciplines and fields of scholarship, this book explores the conceptual and practical aspects of communal ownership of transport. It will be a valuable resource for those seeking innovative approaches to addressing the pressing problems of transport, including graduate and postgraduate students, as well as policymakers, practitioners and community groups.

Leigh Glover is the former Director of the Australasian Centre for the Governance and Management of Urban Transport (GAMUT) at the University of Melbourne, Australia.

Transport and Mobility
Edited by Richard Knowles and Markus Hesse

The inception of this series marks a major resurgence of geographical research into transport and mobility. Reflecting the dynamic relationships between socio-spatial behaviour and change, it acts as a forum for cutting-edge research into transport and mobility, and for innovative and decisive debates on the formulation and repercussions of transport policy making.

For a full list of titles in this series, please visit www.routledge.com/Transport-and-Mobility/book-series/ASHSER-1188

Port-City Interplays in China
James Jixian Wang

Institutional Challenges to Intermodal Transport and Logistics
Governance in Port Regionalisation and Hinterland Integration
Jason Monios

Sustainable Railway Futures
Issues and Challenges
Becky P.Y. Loo and Claude Comtois

Mobility Patterns and Urban Structure
Paulo Pinho and Cecília Silva

Intermodal Freight Terminals
A Life Cycle Governance Framework
Jason Monios and Rickard Bergqvist

Railway Deregulation in Sweden
Dismantling a Monopoly
Gunnar Alexandersson and Staffan Hulten

Community-Owned Transport
Leigh Glover

Geographies of Transport and Mobility
Prospects and Challenges in an Age of Climate Change
Stewart Barr, Jan Prillwitz and Gareth Shaw

Community-Owned Transport

Leigh Glover

LONDON AND NEW YORK

First published 2017 by Routledge

2 Park Square, Milton Park, Abingdon, Oxfordshire OX14 4RN
52 Vanderbilt Avenue, New York, NY 10017

Routledge is an imprint of the Taylor & Francis Group, an informa business

First issued in paperback 2018

British Library Cataloguing in Publication Data
A catalogue record for this book is available from the British Library

Library of Congress Cataloging in Publication Data
Names: Glover, Leigh, author.
Title: Community-owned transport/Leigh Glover.
Description: Abingdon, Oxon; New York, NY: Routledge, 2016. |
Series: Transport and mobility | Includes bibliographical references and index.
Identifiers: LCCN 2016027090 | ISBN 9781472433800 (hardback) |
ISBN 9781315573021 (ebook)
Subjects: LCSH: Urban transportation. | Local transit. | Cooperative societies. | Public-private sector cooperation. | Community development.
Classification: LCC HE305 .G59 2016 | DDC 388.4/042–dc23
LC record available at https://lccn.loc.gov/2016027090

ISBN: 978-1-4724-3380-0 (hbk)
ISBN: 978-0-367-13883-7 (pbk)

Typeset in Times New Roman
by Deanta Global Publishing Services, Chennai, India

For Annie

Contents

Preface

From the classical philosophers to contemporary social scientists and social commentators, freedom has been a major topic for contemplation, aspiration and foundation of political and religious values. Freedom is tied to ideals that are held high in contemporary society, such as self-determination, individual expression, creativity and autonomy. Freedom features strongly in national anthems, national constitutions, party political aims and in revolutionary political works. In contrast, passenger transport is a mundane idea; it is the dull means to desirable ends – the freedom to change location. While freedom is deemed an ideal worthy of seeking and protecting by fighting wars, forging revolutions and a rallying cry for great self and national sacrifice, it would be ludicrous to suggest that urban transport carries anything like the same meaning. Freedom is an inescapably political concept but transport is just a neutral technical system without political values or symbolic importance. Transport is just about getting from A to B. Or is it? Without urban transport, there is no freedom of mobility. All the political debate and contemplation about freedom in an abstract sense can be applied to urban transport. Political values, choices, public institutions, rights of individuals, equality and equity, justice and the limits of authorities and exercise of power are all issues relevant to urban transport. A family standing at the roadside by a lonely sign in the hot sunshine waiting for a city bus whilst private motorcars stream past in their thousands in air-conditioned comfort is an outcome of politics. In the developed world, we have been very good at talking in the abstract about freedom and very poor at thinking about the politics of urban mobility or considering the political causes and implications of urban mobility.

This work concerns a novel idea for providing urban passenger mobility, that of community ownership. There is nothing new about communal ownership; indeed, collective possession by a community or other social group is as ancient as individual ownership and certainly older than the relatively modern concepts of state or corporate ownership. What is novel is considering community ownership models for widespread application and commonplace use in contemporary urban transport. While there have been interesting examples of communal ownership of transport, it has not, to my knowledge, been promoted as an idea with wider potential application.

Although in contemporary times academic scholarship avoids declaring any polemical intentions, at times and places in this work, ideas of polemical character have emerged. Hopefully, it is obvious where this happens and the reader can take it into account. This has occurred for a couple of reasons. First, one rationale for scholarship is to make a difference for the better; there is nothing wrong ethically in pursuing knowledge for its own sake, but here I yield to a higher calling, which is to pursue knowledge for social and environmental good. Second, and following the first point, this volume seeks to change practices and help the transport sector in the transition to environmentally sustainable mobility and wider and more equitable access to mobility within communities.

Placing this work into the existing transport and urban sustainability/urban politics literature and scholarship is not easy, as is the case for many unconventional concepts. Community ownership of transport is a maverick contribution, so it is unsurprising that it cuts across established disciplinary lines of scholarship and professional practices. Urban passenger transport is a field dominated by engineering and economics, giving rise to a particular set of strengths and weaknesses, and successes and failures. A characteristic weakness stems from the narrowness of these disciplinary interests: as a field, problems are interpreted as needing technological and economic solutions. When solutions are not forthcoming from these disciplines, there is a tendency to blame 'politics' and 'human nature' that are deemed to have got in the way of applying appropriate technocratic cures. Either that, or urban passenger transport is declared to be a 'wicked problem' for which no known cures are available.

Alternative ideas and concepts from the social sciences that engage a wider or interdisciplinary approach to urban passenger transport are based on the proposition that there are limits to the conventional approaches to urban transport problems and that unconventional approaches may be beneficial. Although the transport and urban studies arenas have come up with comparatively few new 'ideas' books in recent times, there have been several works that have explored new areas to good effect. Such 'ideas' books include Mathew Paterson's *Automobile Politics* (Paterson 2007), Wolfgang Sach's *For the Love of an Automobile* (Sachs 1984), and David Metz's *Limits to Travel* (Metz 2008). Interestingly, many of these works are by authors whose interests are not primarily in transport and have brought other disciplinary insights to mobility subjects.

Even the emerging field of sustainable urban transport is afflicted by this problem. There is a tendency in most works of sustainable transport and urban sustainability to begin with a couple of wide concept chapters covering what has been known for the last decades or so, followed by a large number of best-practice exemplars. While this model's strength is that technology and (neo-classical) economics are universal and can in theory be applied universally, its weakness is that successfully applying such solutions is almost entirely determined by local social, cultural and geographic factors. Furthermore, given the diversity of experiences and urban circumstances, it is far more difficult to draw lessons from local examples for applying elsewhere than seems to be widely appreciated. Without taking these 'other factors' into account, transformative change through narrow technocratic solutions does not occur. This problem is symptomatic of the lack of attention given to the institutional and political aspects of urban transport. So in contrast to prevailing trends, this is a book about an idea and one that considers the essential barriers to sustainable mobility to be political and institutional, and not economic and technological. It is also a work that seeks to find a place within a number of contemporary social trends and developments and to work out what influences transport and vice versa, rather than treating transport as being at the end of a chain of other demands.

This work is born of frustration with much of the contemporary urban transport discourse and, most certainly, its practices. A few conference papers were presented in Australia on some of the basic ideas and concepts of community-owned transport and these subsequently served as the foundations for this volume. It would be gratifying to write that the concept received encouragement and support from these events, but in large measure, that was not the case as the ideas proposed seemed quite polarising. Tellingly, the negative and disinterested responses tended to come from experienced transport researchers, academics and senior transport officials and more positive receptions came from those with wider urban interests and from outside the traditional transport disciples of engineering and economics. This small insight indicates of much of the state of play in transport research and practice regarding the minor place of the social sciences, a disinterest in issues of governance and institutions and the antipathy towards views contrary to the prevailing neoliberalism in public policy in the Western world.

This work has been a while in the making; when working on the conference papers that were the genesis of this project there were no plans for this volume. It was when I held the position of Director of the Australasian Centre for the Governance and Management of Urban Transport (GAMUT) at the University of Melbourne that I presented these papers. GAMUT was one of a number of Centres of Excellence around the world supported by the Volvo Research and Education Foundations (VREF) that enabled the completion of a major research programme. I remain grateful to the VREF and to my GAMUT colleagues for the opportunities that GAMUT afforded me. A feature of this experience was the opportunity to meet and talk widely with many leading transport scholars in the VREF 'family' and in particular, I would like to thank some of the Directors of other VREF Centres: Harry Dimitriou (University College London), Eliot Sclar (Columbia University), Roger Behrens (University of Cape Town) and Dinesh Mohan and Gitam Towari (both of the Indian Institute of Technology in Delhi). I also had the fortunate opportunity to work with Mark Zuidgeest (University of Cape Town). To protect the interests of all, I should point out that this project on community-owned transport was very much a personal interest, a side-project as it were, and it was not a formal part of the GAMUT research programme and did not involve any GAMUT members.

I should also like to acknowledge my extended GAMUT colleagues over this period and thank them for their support: Kate Alder, Paul Barter, Matthew Burke, Carey Curtis, Jennifer Day, Pan Haixiao, Damon Honnery, Anthony Kent, Crystal Legacy, the late Paul Mees, Fujio Mizuoka, Patrick Moriarty, Kevin O'Connor, Tim Petersen, Bill Russell, Julie Rudner, Jan Scheurer, Sun Sheng Han, John Stone, Sophie Sturup, Patrick Sunter, Izumi Takeda, Carolyn Whitzman and

Marcus Wigan. In particular, I would like to single out my friend, colleague and the founder of GAMUT, Nick Low. John Byrne from the University of Delaware has provided support in many ways over many years.

My Swedish colleague and old mate, Mikael Granberg, gave invaluable assistance to this project and his thoughts were particularly insightful; I am very grateful for his help.

I would also like to acknowledge the efforts and assistance of Katy Crossan and Carolyn Court of Ashgate; thank you both very much. Publication of this work was so ably handled by the Routledge editorial team of Faye Leerink, Emma Tyce and Priscilla Corbett, and with whom it was a genuine pleasure to work. Michael Paye, of Deanta Publishing Services, provided wonderful assistance through his editorial efforts. Comments from two knowledgeable reviewers were extremely helpful and both are thanked for their advice.

Views in this volume do not represent those of any of the aforementioned.

Introduction

Moving to a New Transport Paradigm

There are moments in the course of events when it is possible to look back into history and to contemplate contemporary trends and realise that that you are living in a time of change. Living in a modern and modernising world, you might say that change is the only constant, but this is not exactly true at the scale of societies, for it is the major changes that are important to us. Urban transport seems now to be at one of those inflexions in history; simply put, the remainder of this century will not resemble the last. Looking through older works on transport there is a strong triumphal theme and a narrative of progress – of roads that became larger, of faster cars, of greater freight loads, of more access to mobility, of urban growth and prosperity and increasing road safety. Undoubtedly, the achievements of urban transport have been impressive and transport is, as so often claimed, essential in the modern economy. In the mass consumption economies of the developed world, few other consumer goods have had the cultural impacts of the motorcar and few others have been so invested with symbolic meaning. According to its own narrative, transport is the means of progress, a symbol of progress and an embodiment of progress as it has changed and evolved.

But just as the idea of progress has become a troubled topic in the developed world, so too has urban transport. Within the triumphs of the past are also the costs of transport and it is these costs that have caused the narrative of progress to become undone. Under the rubric of sustainable development, the costs of the modern model of urban transport have begun to be reckoned. High levels of urban pollution, millions of lives lost and ruined through road trauma, the contribution of motorised transport to global warming, the failures to provide mobility for the world's poor, the gross inequities in most nations in access to transport services, the role of transport in increasing social and economic disadvantage and the loss to urban amenity through road building and heavy traffic. Giving a high priority to the needs of motorised private transport in urban planning and design has created cities that have concentrated and reinforced these and other social and environmental ills. For all its achievements, our model of urban transport is helping anthropocentric global warming, ruining lives and communities, diminishing human health and making cities dangerous and unpleasant.

Although there may be constant change in urban transport, we are living in a period of sea change, a shift in paradigm, if you prefer, and to put a name to this change is to call it the rise of sustainable transport. It is the shift away from automobility, based on the ubiquitous use of private cars, and towards sustainable transport, which aims to provide mobility in ways less harmful to society and the environment. Looking back over the twentieth century, the story of personal mobility in developed nations has been dominated by individual travel by motorists using their own cars, powered by fossil fuels and a state apparatus to provide public funding and essential institutional support to secure and promote this model. Developing nations have aspired to this model, although up until recently, only a privileged minority could enjoy it. Today, this model of automobility is largely bankrupt, having failed the goals of social justice (such as the problems of road trauma, inequity in access, iniquitous public subsidy and the geopolitics of the global oil trade), rational resource use (such as the problems of peak oil and having a large ecological footprint) and environmental protection (such as the problems of air pollution, greenhouse gas emissions, losses of urban amenity and car-oriented urban design).

Environmental Constraints

A key reason why future urban transport will not continue the model of mass private transport using privately owned motor vehicles which use energy from fossil fuels, propelled by internal combustion, is due to a set of environmental and resource constraints (covered in further detail in Chapter 3). These environmental constraints begin with the problem of energy for transport, as it has an unusual reliance on energy from oil (in comparison to other sectors of the economy) and it is highly likely that global oil consumption of conventional oil now exceeds that which remains in the oil reserves. By implication, as supply of this globally traded resource declines, its price will rise, bringing an end to the era of low-cost fuel. Alternatives to oil (and oil from unconventional sources) become economically viable, but at higher prices relative to conventional oil prices, so the cost of transport energy will invariably rise. Many alternative fuel and energy sources have been suggested and tried, notably electrification, but all face major barriers to widespread adoption in the short term. Then there are the problems of pollution from transport. International commitments to reducing future greenhouse gas emissions means that fossil fuel use by the transport sector will have to be curbed. Globally, addressing the sector's greenhouse gas emissions may be partially addressed through higher efficiency of fuel use, but such reductions are undermined by increases in the size of the vehicle fleet.

In addition to greenhouse gas emissions, motor vehicles are responsible for a major share of urban pollutants that impose major health impacts. Although a lesser concern, the consumption of resources and energy in the manufacture and maintenance of the world's motor vehicle fleet also represents a considerable cost to the environment. Road trauma is a major cost of motor vehicle use, especially in developing nations. High levels of motor vehicle use produce poor urban environments, either where large road infrastructure reduces urban amenity for the sake of vehicle use or where road traffic is so high that major and minor streets are in almost permanent gridlock.

As developing and newly industrialising nations experience economic growth, the rate of private motor vehicle ownership and use increases. Car and motorcycle ownership rates rise as national income increases, with ownership becoming available to those entering the middle classes and to small private enterprises. Accordingly, there arises the prospect of the levels of car ownership experienced in the developed world becoming the norm in a great many nations in the coming decades. Such a prospect proffers an increase in the local and global costs of transport on the environment and urban societies. Transport planners and city officials increasingly recognise that mass motorisation may be associated with increasing private wealth, but also symptomatic of public underinvestment in mass public transport. Not only is there private wealth accumulation through public subsidy, there is also a transfer of the costs of vehicle use from motor vehicle users onto the environment and other community members, something that is becoming increasingly less acceptable to governments and the wider community. Infrastructure for motor vehicles is another form of private gain and public subsidy, as it reduces both urban amenity and diverts funds that could be used for public transport and walking and cycling infrastructure.

Those Left Behind

Although the development of transport infrastructure in public transport, roads and freeways represented the networked urban transport system, assumptions that this model provides the optimum distribution and allocation of mobility have proven to be flawed. Local and historical forces always shaped the development of these systems, reflecting political and economic influences. Urban infrastructure was the outcome of policies to distribute mobility and this reflected the exercise of political power. What resulted was a fragmentation of mobility opportunities, most evident where public money was spent, with poorer communities receiving less investment than those better off or those locations where there was significant investment, such as factories that required access for employees and logistical needs. As a result, transport infrastructure became part of the mechanism for economic growth; places with higher mobility became more valuable and attracted other investments, thereby further increasing social and economic inequality across cities.

Many inhabitants of developed world cities, including the most prosperous in the world, have never been well connected to transport infrastructure and services and this is even truer of developing world cities. As transport networks have acquired the characteristics of distributed systems and are subject to greater influence of neoliberal philosophies of governing, there are risks of a vicious spiral of under-servicing: little investment in transport followed by low land values and limited economic growth and investment, resulting in lower productivity and socioeconomic occupation that creates economic disincentives for further transport investment. Cities that have concentrated transport funding on road infrastructure create mobility systems with highly uneven mobility opportunities. Households with poor transport opportunities have to bear proportionally more of their own transport costs, such as through car dependency, informal car sharing and informal transport, sometimes of dubious legality or known illegality.

A variant of the problem of exclusion is neglect or under-servicing occurring in locations and for transport services deemed unworthy of further investment. Such is the strength and pervasiveness of the assumption that places of relatively low demand for transport services are 'uneconomic' to service, that rarely do governments, public authorities and corporations bother to express this view in formal terms. Communities, including those with a severe lack of access to transport services have often come to share this outlook. Sometimes, public transport services are simply withdrawn using economic rationalisations. Under neoliberalism, transport planning and development can be influenced by the attractiveness of new projects to private investors, a priority that may well not align with social justice objectives in providing mobility. In fact, conventional economic reasoning suggests that areas of low socioeconomic status are less likely to attract new public transport projects as they may offer lower rates of profit. What can happen, and has happened routinely through the era of motorisation, is that public investment in roads has resulted in neglect of public transport and in competition with new roads, suffers further loss of patronage. Nutley (1996, 2003) observed of rural communities in places with high car ownership that governments failed to recognise any instances of transport disadvantage, were disinclined to subsidise public transport and that those without car access were politically invisible; much the same could be said of those in urban settings (as discussed elsewhere in this volume). Scale can be an important influence on transport investment decisions; governments and private investors may be more interested in providing services to new large shopping centres and areas of high capital investment than in servicing residential areas with dispersed demand with public transport. There is also the phenomenon of 'bypass' that creates segmented and distinct transport infrastructure to service international needs, connecting cities with international logistics, often using high-quality and separate infrastructure from the existing networks; examples include high-speed airport rail links and freeways.

Poor access to transport services is not a new problem, but the growth of cities through population increases and ongoing urbanisation means that the problem of mobility equity will continue to be a factor in future transport policies and practices. Governments with commitments to addressing mobility inequalities will find no satisfaction through recourse to increasing use of private motor vehicles. Without doubt, market forces left to their own devices will continue to promote private vehicle ownership and use, but the viability of this path depends greatly on public investment in road infrastructure. These same forces are largely indifferent to the social effects of large road infrastructure that divides and segregates communities and isolates urban functions, dividing the urban fabric almost effectively as walls.

Rise of Distributed Systems

We are also coming to view transport systems in a different light, particularly the design, use, and financing of transport infrastructure. Over a century or so until the mid-twentieth century, urban infrastructure in developed nations was made up of small islands of services run by regulated monopolies, using standardised infrastructure to provide reliable services in water supply, wastewater, electricity, gas, transport, communications and other requirements of metropolitan living. These networked services grew rapidly in scale and scope, fostered by technological innovation, spreading out across cities, into regions and nations, and internationally and globally. Centralised urban transport systems became integral to industrial growth and central to the wider project of modernity (e.g. Carter, I. 2001, Beaumont and Freeman 2007a, Urry 2007, Schipper 2008).

For reasons of technological, economic and social origins, there has been a shift away from the single model of centralised infrastructure and urban systems towards more disaggregated, poly-centric and varied approaches to providing large-scale urban services, known as distributed systems. Many interpretations of distributed systems are available, partly because the same phenomenon has developed in different fields and in a variety of forms and has been described and understood differently across the academic disciplines. Earlier scholarship that detected the rise of distributed systems, such as Manuel Castells' multiple volume work, *The Information Society*, described how authority, resources and communication has become more dispersed in government, society, politics, economy, culture, knowledge and a range of infrastructure systems (these ideas have origins in earlier critiques of industrial modernity, such as by Georg Simmel) (Castells 1996).

Distributed systems are those in which control and organisation are not centralised and hierarchical but have authority, resources and communication dispersed and controlled, managed by an array of people across an entire system. Exemplars include computing systems that draw on many computers arranged in a computing network, the Internet and electricity generation systems that generate and distribute energy using many sources and forms of generation connected through an extensive grid. Certainly, aspects of these changes are evident in urban infrastructure and city planning, including urban transport systems.

In many urban transport systems there has been a breakdown of the models of single authorities and centralised systems and the rise of more diverse, deregulated and networked transport systems. A clear indication that transport systems are increasingly viewed and operated as distributed systems has been the growing support for integration in transport systems (e.g. Giaoutzi and Nijkamp 2008, Givoni and Banister 2010, UN-Habitat 2013). Integration is applied in many ways, including urban transport operations where different modes and operators run to coordinated timetables and share common ticketing and passenger information systems. Transport infrastructure can also play a role in integration, with passenger transfers between different modes and operators at multi-modal stations and stops, with the concept of 'seamless journeys' across different modes and sub-systems; operators may also share common infrastructure. Goals to achieve 'network effects' in urban transport are consistent with viewing the various components, operations, services and infrastructure as constituting one large system but comprising many different elements. While passenger transport is in the business of distributing travellers, urban transport is itself a distributed system.

Service providers and consumers can benefit from the shift towards distributed systems, with a central theme being a wider consideration of the ways in which services are delivered. Distributed systems are an evolution that reflects both changes in production and consumption, with markets and services becoming more segmented. This can involve breaking up monopolies or unbundling services within a monopoly – an unbundling that can be vertical, horizontal or virtual (Graham and Marvin 2001). Basic economic assumptions that monopolies always have the lowest costs have proved false, as markets are not homogenous and there are segments and niches where other suppliers can meet market needs competitively.

These changes reflect many factors and certainly not just information and communication technologies, although these play a major role. There has become greater differentiation amongst consumers and their particular needs and consumers have become less satisfied with standardised goods, influencing markets to shift away from homogenisation of goods and services and towards customisation. Changing demands over time and space also challenge standardised systems, such as peak periods in road and public transport use, prompting managers to seek ways to shift demand patterns. Entrepreneurs, new specialists and experts are now interposed between producers and consumers and promote further specialisation, especially in demand management.

Distributed systems have both technological and social causes. New low-cost technologies are creating new ways of meeting demands, such as mobile phone applications for carsharing (a concept addressed in Chapter 4). This change is partly prompted by the usually high costs of infrastructure expansion of large systems to meet increased demand (especially in transport infrastructure). There are also public policy changes shaping the circumstances favouring distributed systems. Reducing absolute demand or flattening out 'peaky' demand patterns through demand management strategies has often proved cheaper than enhancing supply. Distributed systems offer a range of ways of dealing with demand, as integrated transport networks provide a wider range of choices and options than can be achieved under traditional models.

Views differ on the merits of this transformation. Graham and Marvin (2001) describe the changes in UK urban infrastructure regulation and management as part of a political process as a splintering of urban networks (covering their provision, development and control) that results in a fracturing and dispersal of institutions, the result of which is increasing social inequity. Those who view these changes as part of economic reform, such as the Bretton Woods institutions (notably the World Bank and the International Monetary Fund) and multilateral development banks (e.g. the Asian Development Bank and the European Investment Bank), who support such processes. Changes from centralised systems to distributed systems may be associated with greater involvement of corporations in public enterprises and privatisation, but there are also opportunities (at least in theory) for other services and participants, including community organisations.

Technological Change

Socially and economically there are other aspects of the era of change in which we live, with changes in politics, corporate activities and government institutions that are also changing urban transport. These developments have prompted technological change and, in turn, there have been technological developments that have fostered new ways of operating and managing transport systems. Many classifications of transport technology developments are available and it is, in some senses, a baffling array. Great attention and resources have been devoted to motor vehicle technology, including their fuel/energy sources (an issue considered in Chapter 3). There are also a set of new technologies and applications using digital and information technologies, including those for transport system management (covering such activities as congestion charging, access charging, traffic management, public transport vehicle management, journey planning and autonomous vehicles) and those facilitating the so-called 'new mobility' activities (such as car sharing, bicycle sharing, lift sharing, real-time information systems on roads and parking).

While the effects of the new communication and information technologies can be exaggerated, the digital revolution has changed some of the established ways in which urban transport operates and some of these are likely to prove profound. Real-time information about services is now available, so that public transport operators can manage their systems more effectively and this improves the opportunities for system integration and coordination. Passengers also reap the benefits from this information, as information systems and on-board vehicle telemetry provides advice on waiting times, service availability, travel options and estimated travel times. Similar advice can be used by road transport managers and operators for planning and for operations to manage traffic, enable pay-for-use schemes and congestion charging, access charges, and other functions. Vehicle manufacturers are using new data technologies within vehicles in a host of ways, including for driver advice, vehicle control, engine management and other functions, with the attention-grabbing prospect of driverless cars.

It is worth remembering that perceptions of mobility change over time, particularly when compared between eras with different technologies and transport choices. In the developed world, the advent of low-cost airlines has made formerly exclusive air travel more widely accessible, with international travel and long-distance trips becoming quite commonplace for the middle classes. Where car ownership is common, urban trips of, say 10 kilometres, are typically regarded as routine and something that does not require much effort, expenditure or time. Such perceptions are entirely different from those eras before motorised transport or for those without car access. Settlements 10 kilometres from large cities in places and times where most people had to walk or take a horse-drawn vehicle were considered to be at some distance. When the railway towns and new suburbs were developed with the arrival of the railways and electric trams they represented an escape from the inhospitable aspects of the industrial cities, but often they were but a few kilometres from the large cities. We inhabitants of the modern world in developed nations have become very casual, if not dismissive, of what we regard as short trips, but anyone who has returned to a reliance on walking will soon re-calibrate their understanding of a short trip. Putting transport technology into perspective can be difficult; its power to transform mobility and our perceptions of travel is profound and this can lead to overly optimistic views as to the potential of technology to address all major transport problems.

Neoliberalism

Linked to many of these changes is the underpinning political philosophy that has dominated the thinking and public policies of developed nations and many developing national governments and key international agencies (such as the aforementioned World Bank, International Monetary Fund and the regional development banks): that of neoliberalism. Politically, most of the English-speaking world and a good number of the other OECD nations have political agendas dominated by neoliberal approaches. For transport, this has seen many of the large state transport bureaucracies broken up, with services formerly provided by public servants and state agencies now provided by contracted private firms, and a greater interest in market-based approaches for public policy and less interest in direct regulatory control. This has given rise to a multitude of new practices, from the way new transport infrastructure is financed, to the outsourcing of maintenance operations on transport, to franchising public transport operations, to the use of congestion charging, access charging and more pay-for-use services. Nearly all road and public transport infrastructure and other transport capital around the world up until relatively recently was funded by government bodies, usually under models of the welfare state in various guises. Such models are no longer universal and this change has been fundamental in the rise of a new transport paradigm.

Whatever or not we approve of, or reject, neoliberalism, it is clearly the dominant political ethos of the developed world and appears to be similarly taken up in the governance of many developing nations. Neoliberalism's exact meaning is debated and although there is philosophical ambiguity (over such points as to whether it is laissez-faire economics or not), there is strong consensus on what constitutes neoliberal economic practices by OECD governments, so it need not concern us here whether or not the practices and ideology of neoliberalism are neatly aligned. OECD governments have not adopted the neoliberalism programme in a pure form, as this would not be possible in liberal-democratic states, but have inclined policy towards that end. Key aspects of neoliberalism in theory are that governments should only have a minimal role in society, particularly in economic matters, allowing that governments should protect the nation state and ensure the rule of law so that markets are protected from without and from within (to prevent abuses of markets, such as the rise of monopoly corporations). Neoliberalism holds that markets should perfectly reflect individual and community values and preferences, and therefore questions the legitimacy of governments acting to express community values or to initiate policies and programmes of a progressive character. In essence, neoliberalism only supports those values that enable the free market exchange of goods and services within an economic setting. Insofar as values do not impinge on such market exchanges, neoliberalism does not have a position on the expression of other values (and thereby differs from classical liberalism's support of values such as reason, tolerance and justice). In pure from, neoliberalism is *economic liberalism*, which is centred on the rights of individuals and their freedom of choice within economic markets and that, as far as possible, markets should be self-regulating (i.e. functioning free of the coercive power of governments).

It has been proposed that conventional left–right political divisions that marked most of the political divides in the representative governments of the developed world through the last century have weakened and, between these polarities, as it were, has arisen a 'third way' (Giddens 1998). There are considerable differences, if not confusion, over what this third way stands for in ideological terms and it appears to combine the progressive agenda for social issues and a neoliberal agenda for economic issues. Another strand of political observation has been that both conservative (read as right wing) and progressive (read as left wing) political parties have become more centralised, less ideologically driven and more pragmatic. Debate over the place of formalised centrist political parties in the developed world continues, but within the broad sweep, a reasonable generalisation is that the major left wing parties have shifted to the right on economic policy.

Although this shift to neoliberalism may be debated in political philosophy, the adoption and practice of neoliberal economic policy can be shown empirically in many nations. Consistent with the underlying philosophy of liberalism, there has been a movement towards free trade and open markets, market-based policy instruments, and deregulation and privatisation, all with the goal of reducing government's role in the economy. Neoliberalism has supplanted the dominant practices of the welfare state model, albeit much of the social welfare apparatus (and a considerable amount of corporate welfare as well) has typically remained in place following neoliberal reforms. Global and national economic crises have prompted governments to take up neoliberal economic policies and these are often combined with austerity measures and cutting government expenditure to reduce public sector indebtedness. While such changes are largely within the realm of economic and finance policy, neoliberalism is a political philosophy and what we have witnessed is a change in the political values that underpin economic policy, changes that in turn produce political outcomes through the economic system.

Urban transport, as a major recipient of public funds (both capital spending and recurrent expenditure) has been at the forefront of neoliberal policy reform, so that in many OECD nations the contemporary transport systems are quite different to what existed 25 years ago (e.g. OECD 2003, Macário *et al.* 2007, Farmer 2013) Institutional reform is a major part of these changes under neoliberalism (as discussed in Chapter 2). An obvious change has been replacing public sector ownership and control of public transport with corporatisation and privatisation strategies that have seen management, operations and maintenance pass from public to private realms. Changes in employment practices have been somewhat slower, given that transport workforces have usually been highly unionised, but this too has given way to un-unionised employment, featuring contracts instead of salaries and performance-based management practices. Financing too has been greatly altered; funding for major projects is no longer the exclusive role of government, with new arrangements, such as public–private partnerships, designed to shift both risk and financial burden across to the private sector.

Politically and economically, neoliberalism is bound up in many of the developments in transport described above. Certainly, the trend towards distributed systems in transport networks and the segmentation of urban transport markets is tied into neoliberal economic policies, as are many of changes in public institutions governing transport. Neoliberalism is also bound into those new technologies and technological applications exerting an influence on urban transport and with the rise of business end of collaborative consumption and the sharing economy (the subjects of Chapter 4). No explicit judgement is being offered of neoliberalism here, but there can be little doubt that it has produced new challenges to the access and equity issues in urban transport and to the broader goals of sustainable transport, because progressive measures on these challenges have been traditionally based in public sector regulatory measures and expenditure programmes.

Developments in the Sharing Economy and Shared-Use Transport

Shared-use transport is a rapidly growing component of urban transport and the subject of several innovations in urban mobility, such as the sharing of motor vehicles, lifts in vehicles and bicycles. Whether or not this is the extensive transformation that the share-use promoters are suggesting remains an open question, but without doubt it has had an impact on transport, albeit within a relatively small range of activities. Largely, these new ways of addressing mobility problems result from the use of mobile phones, the Internet and the associated technologies and have been driven by a range of motivations, including environmental awareness and personal budget decisions.

There is a confusing array of terms covering shared-use transport modes and, accordingly, it may be helpful to spell out some of the more common terms, most of which appear in various places in this volume, using those suggested by Shaheen and Christensen (2014, see Table I.1).

Is there Another Way?

Many of these aforementioned trends that are shaping the future of urban passenger transport have existed all the way through the era of modern urban transport (i.e. transport of the industrial era), but it is in contemporary times that all have taken the particular form that makes each highly influential. Critically, these

Table I.1 Terms in shared-use transport

Carsharing	Round-trip/Classic	Car fleet used for round trips that require users to pay by the hour and/or/ mile.
	Peer-to-Peer	Shared use of private vehicles, typically managed by third party.
	One-way	An operator-owned car fleet used for point-to-point trips, facilitated through street parking agreements; payment usually by the minute.
	Fractional Ownership	Individuals sublease or subscribe to a vehicle by a third party and share use of that vehicle.
Bikesharing	Public	A fleet of bicycles primarily for point-to-point trips that require users to pay a fee in order to receive an unlimited number of rides within a time interval.
	Peer-to-Peer	A bicycle that is rented or borrowed from individuals via a mobile/web application or bike rental shop.
	Closed-Community	A fleet of bicycles placed within a campus, business, or institution for private use by campus persons, normally for roundtrip.
Ridesharing	Carpooling	Grouping of travellers into a privately owned vehicle, typically for commuting.
	Vanpooling	Commuters travelling to/from a job centre sharing a ride in a van.
	Classic Real-Time	Grouping of travellers into privately owned vehicles, typically for commuting.
On-demand Ridesharing	Transport Network Companies	A service that allows passengers to connect with and pay drivers who use their personal vehicles for pre-arranged trips facilitated tips facilitated through a mobile application.
	On-demand Professional Driver Services	A service that enables passengers to find, hail, and, in some cases, pay for a professional driver (taxi or black car) through a mobile application in real-time.
Scooter Sharing		An operator-owned fleet of motorised scooters made available for roundtrip or one-way use that usually require users to pay by the hour or minute for use.
Shuttle and Jitney Services		A privately owned bus service for community or commuter transportation purposes.

Source: Shaheen and Christensen (2014).

trends in environmental constraints and environmentalism, inequity in access to mobility, the rise of distributed transport systems, the politics and practices of neoliberalism, technological change and the formalising of the shared economy and collaborative consumption are not all pulling the same direction. For example, neoliberalism and the shared economy are clearly mutually supporting phenomena, whereas neoliberalism and environmental protection are in a far more contentious relationship. Predicting the broad future of urban transport is, therefore, largely impossible because of the diversity and complexity of these trends. It is certain, however, that the urban transport models prevalent in the last century will not persist through this century, as evidenced by the changes already underway.

Scholarship and expert commentary on contemporary transport engages with these trends, but usually within a tightly constrained scope that rarely takes a comprehensive view of these different developments. Work on sustainable transport obviously deals with these environmental issues, but as with the usual approach to urban transport, it is largely taken that the enablers of the transition will be primarily technological and economic. Many of the books and much of the commentary advocating sustainable transport are proscriptive, collating exemplars of good and admirable practices from around the world. While such 'best-practices' are illuminating, if not inspiring, the essential messages concern technocratic solutions divorced from considering the barriers of politics, institutions and the like. Oftentimes, such calls are sheep clothed as wolves; the rhetoric suggests the need for environmental reform, but without considering the need for social change and relying on technologies and market-based solutions, the status quo is unlikely to be disturbed. One of the reasons why technological solutions are eternally popular is that they usually do not directly threaten established institutions and proffer improvements without the need for contentious restrictions of resources use or calls for restraint in consumption. Engineering and economics have long set the discourse for urban transport. Essential disciplines for operating complex and demanding transport agencies and operations, engineering and economics have an essential role in urban transport. But for reforming and changing such systems, as evidenced by the sorts of trends shaping the changes in contemporary transport, the insights, concepts, and practices from a wider array of expertise, understanding and outlooks are necessary.

When we consider the problem of how to implement reforms in the transport systems around the world, the discussion focuses on two approaches: private transport and public transport. Seemingly, all the options and choices are founded in either or both of these options. This book is concerned with a new idea and an alternative approach, a third option, that of community-owned transport. It is an idea that stands against the tide of transport history in many respects. For a start, urban transport has been dominated in the last century by private transport and before that, mass public transport, with community-based collective approaches to transport barely existing. During this era of individualised mass consumption, the idea of owning less and consuming less is revolutionary. In other respects, community-owned transport shares many of the features of the changes underway in the transport sector, such as the rise of collaborative consumption.

Community-owned transport is an innovation and an idea for addressing at least some of the social, economic and environmental failings of automobility.

Community-Owned Transport

In many ways, community-owned transport occupies a unique position and is quite distinct from private transport, such as privately owned cars and motorcycles. It is also distinct from public transport that typically provides a mass service through such modes as subways, railways and buses. However, it also shares a number of characteristics with these more familiar institutions and this gives community-owned transport an ambiguous character if considered only in terms of these more familiar systems. This volume approaches the issue of passenger transport from the perspective of ownership; Table I.2 lists the basic forms that ownership can take.

In terms of scale, by the number of users, community-owned transport sits between the very small scale of individual transport, such as using passenger cars, and the mass scale of public transport, although in most practices caters for relatively small numbers of passengers. It follows that the sort of social institutions involved in these different models of mobility differ greatly, with private transport entailing using private vehicles and public infrastructure with public oversight and management systems (such as vehicle and driver licencing and traffic regulation over vehicle use). Public transport is controlled and managed by a largely different set of institutions operating under a range of models, from state ownership and operation to franchised private operators under government contracts. Community-owned transport operates largely under less formal institutions, being based on a variety of collective organisations and agreements.

Autonomy of each traveller is another source of difference between these mobility choices. In this volume, we are particularly interested in the implications of mobility choices for individuals and households and the social, economic and environmental consequences of these choices. Along a continuum of autonomy,

Table I.2 Types of transport ownership

Form of Ownership	*Owners*
Private	Individuals Firms Franchisees with state contracts
Public	Government (national, state, local, city, regional)
Informal	Individuals without legal status Firms without legal status
Charity	Charitable trusts and enterprises Not-for-profit businesses Social enterprises
Community-owned	Community organizations (many forms)

there is a range of freedoms. For individual transport modes, there is a wide freedom of mobility, represented culturally by the notion of 'the freedom of the road'. Here, we will argue that in many respects, individual mobility modes provide only a constrained and highly conditional freedom, a freedom that is determined by having access to a vehicle, the wealth to operate the vehicle, a particular set of skills and so forth. Further, such freedoms are only realised through a vast and complex intervention of governments and markets to furnish the means for individual vehicles to have mobility. Only when all these other conditions are in place does the user of private transport have considerable choice and comfort. Public transport offers mobility choices determined by the design and operation of the system, such as the location of stops and stations, service timetable and opportunities to connect with other services. In some sense, community-owned transport lies in the middle of this range of autonomy, as it offers some flexibility and opportunities for individual choices, but importantly, usually as mediated through collective decisions.

Flexibility over travel, such as when, where and how to travel, is closely tied to autonomy of movement, although they are different concepts. One of the important features of effective mobility within cities is the extent to which the integration of transport systems allows the network effect in mobility. Although transport networks may comprise a dense matrix of stops and stations across a city, it is the degree to which different services connect that determines the functional use of this network. Highly interconnected and coordinated services within a network offer the optimal degree of connectivity and opportunities for journeys of linked trips. Such integration provides the flexibility travellers seek and highly integrated public transport systems can provide attractive alternatives to private automobile trips through the range of destinations and return trips available, minimising waiting and travelling times, and providing a range of choices over the timing, routes and other travel benefits. Community-owned transport will be optimal, therefore, when it adds to the flexibility of travellers through its integration with existing public transport services.

A Few Words on 'Community'

A reasonable challenge to community ownership of transport, or community ownership of anything else for that matter, is to pin down exactly what 'community' means in this context. Several generations ago, when Hillary (1955) famously found 94 different meanings to the use of 'community', it was clear that the concept had wide and varying usage and no doubt, this tally of meanings has only continued to increase. Community covers meanings that relate to places (as part of human ecology), to social structures and institutions, and to symbols, wherein community is conferred through shared cultural identifiers. An older concept that has (re)entered the social science lexicon, namely 'civil society', makes clarifying the concept of community as used in this volume a little easier. Civil society captures the idea of the realm of social life not identified as either corporations or governments; necessarily, this covers a wide array of 'third sector' entities and activities. Significantly, civil society involves voluntary participation and social networks, such as formed around family, interest, ideologies and beliefs of various kinds. Conceptually, this loose definition fits well into the urban transport setting where the activities of governments and corporations are so familiar, so that we know that community-owned transport differs from these familiar models of ownership.

Taylor (2000) makes the point that community can be a descriptive and proscriptive term and suggests that it is used descriptively, normatively and instrumentally. For the purpose of understanding community-owned transport, all three dimensions of the concept are applicable. Certainly, community is given a descriptive connotation because community ownership of transport denotes an entity that is shared between a group of people with a common interest in a mobility service. There may also be other dimensions to this common interest, such as shared social relations arising from a common experience in a particular residential location, common social ties or experiences (or heritage) and participants may be members of other social groups, such as sporting or interest clubs, volunteer organisations and cultural groups. In practice, there are likely to be many overlapping interests and associations, so that individuals and households have numerous and diverse alliances. Although a community may immediately conjure a social group identified with a particular place, the role of place in identifying communities can be contentious. What academics and researchers seek, however, in identifying community, is the role of shared interests.

Community can have a normative interpretation (meaning 'what ought to be' as opposed to 'what is') and this also applies to the use of community ownership in this volume. Invariably, much of what is promoted on enhancing community relationships is premised on assumptions of restoring lost associations, harking back to the effects of the large-scale social change, including those from the Industrial Revolution, urbanisation, mass consumption, industrialisation, capitalism and the welfare state. Promoting communities need not, however, have a retrospective cast and the revival of community policy in Western nations seems to be more concerned with new forms to democratic processes and greater inclusion in public policy processes of marginalised groups as part of the movement for social democracy. These developments were in part a recognition of the limits and effectiveness of the neoliberal agenda in shrinking government and using market-based approaches to social policy and also a result of vigorous efforts by NGOs and community groups to provide services that governments and corporations had neglected.

For community-owned transport to function, the community necessarily has an instrumental role. Efforts seeking great community engagement have primarily come from governments and take a wide variety of forms, with the use of the term community covering an array of services, including a range of consultative and engagement activities, community delivery of services, cooperative activities between community and government organisations and financing community activities. In effect, citizens and citizen groups become agents for public policy formulation, delivery and monitoring. Despite greater interest from many quarters in increasing the opportunities for greater community engagement in public life, there are also suspicions behind the motivations of governments' interest in such engagement. On the one hand, there are clear benefits to strengthening community activity, but on the other hand, such initiatives may be exploitative, ineffectual, cynical and a strategy of the 'hollowed out' public sector to evade its wider responsibilities.

A long-standing institutional form of community organisation is the mutual society or mutual organisation (often just called 'mutuals' in the UK). Mutual societies have a history through the industrial era, covering cooperatives, employee-owned businesses and other forms of enterprise owned by employees, service and product consumers, and suppliers or some combination of these. Prominent in industrial and social class history, the zenith of such groups was the nineteenth century, covering such services as banking, housing, insurance and health care. There was a subsequent decline in the twentieth century and over this period many mutual societies became institutionalised in public entities, corporations under private ownership or more formal associations. Originally, mutual societies played an important role in providing services to working-, and later, middle-classes and followed a programme informed by progressive politics in the UK, US, Europe and elsewhere. Despite this decline, mutual societies remained important in some locations, notably in some European nations, such as Italy and Spain.

Mutual societies have undergone a revival in recent times, in line with the wider promotion of community-based public policy initiatives and self-generated community activities (Noya and Lecamp 1999, Nyssens 2006, Ridley-Duff and Bull 2011). This revival has highlighted the benefits of grounding activity within communities. Families and communities can provide welfare services, in the broadest sense, in places and in ways where governments are ineffective. Conservative critics support such activity on the grounds that state welfare diminishes community capacities in self-reliance and that individuals and communities have a moral obligation to assume responsibility for their welfare. Boosting community ties may also help counter the declining support of formal political institutions, electoral processes and political allegiances prominent in many developed nations, a process that some see as a decline in democracy itself (Putnam 2001). Certainly, collective behaviour enhances social cohesion, cooperative activities, promotes personal relationships and provides means to create equitable access and use of local resources. On the other political side, some support greater community activity as a counter and bulwark against neoliberalism and the decline of the welfare state. Community activity is built on and strengthens the role of trust and reciprocity, features that supporters of social networks and social capital consider as essential.

Advocacy of community activity, as explored and supported in this volume, needs to consider the criticisms and limitations levelled at it. Given that community supporters have drawn on earlier, pre-industrial times, it can be said that this is a romantic view of traditional and rural societies. Communities can be formed by the conscious exclusion of 'others' against whom prejudices are harboured; communities can also give rise to fears and prejudices that foster injustices against, and conflicts with, those outside the community. Such dynamics can also apply to community members from whom conformity is one price of membership, leading to societies marked by conservatism, hierarchy and rigidity. Calls of support for community by governments and political leaders can constitute hollow propaganda, by presuming that communities are bound by an unspoken political ethos; actual differences in power and existing conflicts are ignored and left unresolved.

Not to undervalue these criticisms, but many involve questions of 'big politics', such as whether civil society as a whole can bring about wide-scale social reform given its fractured identity and diverse composition. Such questions involve greater problems than that of urban mobility systems being considered in this volume. Critically, there is a movement in place promoting greater community engagement and community-owned transport is placed firmly in this trend. Nonetheless, this volume does consider the strengths and weaknesses of community ownership in addressing key urban transport problems.

About this Book

This work seeks to:

- develop and identify the concept and potential practices of community-owned transport;
- see how this idea might work in practice;
- consider how community-owned transport could help bring greater environmental sustainability to the urban passenger transport sector.

However, this book does not set out to tie down the community-owned transport sector by dictating what might qualify as community ownership and what does not, in some definitive manner. This reluctance stems from the idea that an effort to tightly define or categorise community-owned transport would be to limit its potential from the outset.

There is a revolution underway in passenger transport as the sector shifts away from providing transport services based around vehicles and transport infrastructure to providing mobility services in ways consistent with sustainable transport. What does this mean? Traditionally, governments built roads and associated infrastructure for people and firms to use their own motor vehicles and developed a considerable institutional capacity to facilitate private vehicle use, covering laws and regulations over the road vehicle design and use, road use, financing of public infrastructure, policing and regulation of drivers and vehicles, and town and urban planning for vehicle use. Critically, the emphasis was on facilitating activity by private users. Public transport has traditionally had state authorities (or state-regulated private entities) and provision vehicles, rolling stock and infrastructure and an organised system with fares, timetables, service standards and the like. In sum, the idea for transport is that the state provides the basic means for passengers to undertake journeys, but not much more.

An obvious limitation to this traditional approach is that, in large measure, people do not travel in cities as a pleasure unto itself – rather, transport is a means to another end; we travel to achieve some other purpose, such as to travel to work, to shops, to school and to visit people. What travellers seek from transportation is access to another location and what they value is mobility (that can be reflected in measures of time and money). Success in such systems was often measured in terms of mobility, using metrics of people movements. From the perspectives of transport system users, such an approach has an obvious limitation; people care about whether the service meets their specific needs, not just whether a service was provided. This is the difference between measures, such as passenger-kilometres of services (of interest to transport system providers), and measures such as overall journey travel time as experienced by individual travellers. Accordingly, accessibility measures are now used to assess various aspects of transport system performance. These measures are broader than just transport, as access to mobility can involve aspects of urban and land use planning by placing trip origins and destinations in closer proximity. Access also evokes questions of equity in mobility provision, as assessed by the availability of services to those who are transport disadvantaged. Accessibility can also include access by means other than transport, such as by substituting transport by using telecommunications technology.

This volume seeks to find the place for community-owned transport within the emerging paradigm of sustainable transport. It is not intended to be a polemic, although in attempting to put the case for recognising and expanding a neglected realm of urban transport systems, doubtless some aspects will be polemical in seeking to promote change. Part of the motivation behind this work is born of a frustration with the narrowness of much of the contemporary debate over urban transport and intransigence among the majority of urban transport academics, researchers and practitioners. Such intransigence covers several aspects of transport policy and planning:

- A determination to ensure that only qualified specialists in the field are able to contribute to its planning and decision-making in the public and private spheres (and an associated reliance on engineering and economic disciplines).
- A general disinterest in access and equity to urban mobility as problems for the transport sector and a willingness to hive off responsibilities and the search for solutions to such problems to the welfare sector and to affected individuals.
- An outlook that holds mobility as strictly a derived demand, for which the causes and consequences are responsibilities for other stakeholders or are issues created by society in general and abstract societal trends.
- A siege mentality regarding the transition to sustainable transport that finds those transport organisations and key stakeholders adopting an outward

position of claiming to be waiting for others to deliver the means necessary to enable the transitions needed (while internally doing little) in order to maintain wider credibility.

- A passivity, if not welcoming, of the constant slide towards greater neoliberalism in the setting of policies and practices within the sector and creating a milieu that treats such changes as neutral and technical innovations, without thoroughly evaluating such measures or revealing openly the political values being endorsed by such practices.

Community-owned transport cannot hope to overturn such problems in the transport sector, but it may have two direct and beneficial effects. Firstly, it can provide a model of practices and innovations to address each and every one of these failings, even if that overall effect may be small. While the eventual mode share of community-owned transport may be minor, it may make a substantial difference to the lives of those individuals that previously had few mobility choices. As a whole, the transport sector could markedly improve its performance in taking access and equity as a major area of responsibility, rather than treating as marginal those suffering mobility disadvantages whose patronage is considered uneconomic to court. Community ownership may also prove to be of some significance in the future, a possibility that should not be entirely dismissed. Secondly, community ownership can serve as the theoretical and conceptual foil to the failings of conventional urban transport systems. Through its existence, community ownership can stand as a critique that exposes the political values, beliefs and expectations of the existing transport systems and those who control it, thereby threatening those with vested interests in the current systems of ownership. Transport knowledge is dominated by economics (of a conventional stripe) and engineering, and this will not be changing soon, especially as 'we are all neoliberal now'. In this book, we want to go beyond such self-imposed limitations and consider a wider range of possibilities; our interest is in revolutionary change rather than the diminishing opportunities of further evolution.

There are four sections following this Introduction:

Part I: Modern Urban Transport and the Challenge of Sustainable Transport

These first three chapters establish the foundations for understanding the basic reasons behind the condition of contemporary urban transport that have created the problem of addressing the challenge of sustainable transport. Chapter 1 provides a history of modern urban transport systems, beginning with the Victorian industrial city as created by the High Industrial Revolution and tracing the development thorough the following stages that established the contours of contemporary urban transport. Chapter 2 examines transport governance and its associated institutions, focusing on the central issues and ideas that explain the rationale for government involvement in urban transport and the institutions developed to undertake the public ownership, planning and management roles in urban transport. Some key issues arising from public sector governance are examined, with an emphasis on their implications for the major issues facing urban transport. Chapter 3 covers what is perhaps the most important of the major challenges to conventional urban transport, namely its high environmental and social costs and the broader issue of sustainable urban transport. Key issues canvassed are greenhouse gas emissions from the transport sector, road trauma and the health impacts of transport, peak oil and the associated energy issues, and access and equity in urban mobility.

Part II: Foundations of Community-Owned Transport

These three chapters expound the broader concept of community ownership, canvassing some of the political theories behind it, looking at urban mobility as an issue of resource use and considers the implications of these ideas for urban transport. Chapter 4 deals with collaborative consumption and the growing interest in 'the sharing economy', which is of particular interest because some of the key enterprises in this emerging and rapidly growing economic sector involve urban transport. Chapter 5 examines the political roots of community and collective ownership based around ecological goals and environmentalism through the concept of 'eco-communalism' and is largely theoretical. Chapter 6 introduces the concept of 'the commons', beginning with an explanation of the concept and then moving to its application to urban mobility.

Part III: Community-Owned Transport Practices

This section shifts the discussion from articulating the problems facing urban transport and the foundations for the alternative approach of community ownership onto the issues associated with the practices of community-owned transport. Chapter 7 draws the themes of Part II together and defines and explains the concept of community-owned transport. This chapter also identifies the place for community ownership in the context of the limits of public and private ownership of transport; it also differentiates between community-owned transport and welfare transport. It also describes a variety of models for community ownership of transport, using both theoretical models, such as social enterprises, and descriptions of actual practices, including some from eco-communities. Chapter 8 returns to the theme of sustainable transport and articulates a number of ways that community ownership of transport can contribute to the goal of sustainable transport, such as by providing an alternative to private car use and encouraging increased use of public transport, walking and cycling. Chapter 9 looks at the possible objections to community-owned transport and considers the counter-arguments, dealing with such issues as fragmenting the transport system, the problem of small-scale inefficiency and the undermining of public transport.

Part IV: Implications of Community-Owned Transport

Two chapters close the volume. Chapter 10 develops some of the political themes of the volume in a discussion of the political economy of community-owned transport around the selected themes of the limits to governments and corporations in providing services, convivial society and neoliberalism. Chapter 11 provides a closing argument for policymakers to consider the benefits of supporting community-owned transport.

Part I

Modern Urban Transport and the Challenge of Sustainable Transport

1 A History of Modern Urban Transport Systems

Today's crises of urban transport, as outlined in the Introduction, are the culmination of many historical trends, some with deep roots in the early motorisation of urban transport while others are comparatively recent issues that have taken time to develop. Insights from the development of urban transport can assist in understanding how modern urban transport systems are governed and how particular institutions are involved in this process, matters examined in the following chapters.

Overview

Cities and their inhabitants' lives were transformed by the Industrial Revolution (1750–1850) that began in the UK; urban transport was also revolutionised and was itself one of the major forces of transformation. Prior to, and for much of the duration of the Industrial Revolution, urban transport was largely as it had always been: un-mechanised and reliant on the corporeal energy from human and animal efforts in walking and harnessed beasts of burden. In contemporary cities of the industrialised world, animal transport has vanished and walking accounts for few trips of significant distances. Now, over 150 years since the Industrial Revolution's end, urban transport in the developed world and increasingly in the developing world is almost entirely mechanised, drawing on external sources of energy, mostly fossil fuels, and comprising mobility systems dominated by mechanised and motorised vehicles. This evolution in transport systems occurred quickly and altered the course of human history. As a consequence, there were fundamental social and cultural changes resulting from how people travelled, why they travelled, where they travelled to and from, and the costs and rewards of such travel. Arguably, modern travel systems also changed at the societal and personal levels, giving rise to new values, perceptions, preferences and consciousness. Urban settlements have also come to reflect these technological and social changes from the evolution of urban transport. There are many fine transport histories available, both of a general kind and of particular places, and this chapter draws on this work, with an emphasis on the United Kingdom and United States because these places are where many of the changes first occurred, changes that subsequently became globalised.

Transport histories often emphasise technological developments, concentrating on infrastructure and vehicles. For convenience, accounts of modern urban transport's evolution are typically divided into phases of technological emergence and development. This is, of course, something of an abstraction that neglects the rivalries between competing technologies and risks, implying that the evolution of urban transport has followed a constant and linear path determined by a set of well-understood causes. Identifying the new transport modes that succeeded is easy in hindsight, but explaining the absolute or relative failing of competing technologies is not always so obvious, nor is forecasting technological change a reliable enterprise.

While following the conventional march of successive transport technologies, this overview of the industrial phase of urban transport history is based on the proposition that this development has been within the sociocultural domain of modernity (a concept explained more fully in the final section of this chapter). Urban transport history varies between locations and times and the resultant systems may share the basic technological elements: typically a car-based system of private transport and public transport system with heavy rail, light rail and buses. There are many local differences in transport system resulting from the local circumstances of history, geography, culture, economics and other factors. There is, however, a single model for modern urban transport and a common pattern of technological evolution towards that model, albeit with many local variations in the speed, extent and exact configuration of those changes. Histories of urban transport have recognised this evolution and the model of urban transport, both as a broad phenomenon and in local histories of settlements, urban transport and specific modes, usually railroads, trams, motor vehicles and bicycles. In this account, it is posited that the broader sweep of transport systems in the modern era have all followed the dictates of modernity, rather than just single modes of transport or particular locations. Explanations of the factors behind this technological evolution are less often explained and despite the existence of modern transport systems around the world, most historical accounts are reluctant to draw many generalisations about the process of technological and urban evolution.

One factor complicating urban transport's evolution is the complexity of the systems; indeed, a feature of these systems is increasing complexity over time as new technologies and sub-systems are added and there is constant experimentation and innovation. While some technologies become redundant, others continue to be used, so that the total mix of technologies and systems increases over time. New technologies can comprehensively displace existing technologies; the intercity steam railway supplanted the horse-drawn coach system fairly rapidly through the mid-eighteenth century; the electrified omnibus took the place of horse-drawn omnibuses with similar rapidity at the end of that century, and there was no turning back from these changes. Old and new technologies can also coexist over the longer term, as evidenced by the array of transport technologies in modern cities; electrified trams and railways, buses, private motor vehicles and bicycles have continued to be used for the better part of 100 years. Associated urban development creates new transport demands, economic development and social change spurs greater specialisation and variety within the economy and society, while urban growth promotes increasing demand for transport, factors that make transport demands more complicated, with greater variety in mobility needs.

Against the turbulent and frequent change of this post-Industrial Revolution period, there has also been a curious constancy in aspects of urban transport. Noticeably, the basic components of transport – a world of roads and rails – have changed relatively little. Someone alive in 1900 would have little idea of the use and purpose of contemporary information and communication technologies, for example, but would easily recognise the transport systems within contemporary transport. As well as our familiarity with trains, trams, motor vehicles and bicycles over the past century or so, the way that we understand and use urban transport has changed, as have the cities in which we live. In developed nations, contemporary perceptions of travel have been formed by mass motorised mobility and this influences our views on speed, distance and time. Today, for those with access to a motor vehicle, employment 10 kilometres from home offers few challenges; before 1900, most trips were pedestrian, something that casts a 10-kilometre trip in an entirely different light.

A key change in cities since the Industrial Revolution is that of scale. Globally, the level of urban population, extent of urbanisation, the number of nations becoming urbanised and the size and number of very large cities is without precedent (Satterwaite 2007). In 1800, only London and Beijing (then Peking) exceeded 1 million inhabitants and 100 years later there were 17 such-sized cities, rising to 75 by 1950 and to 380 by 2000 (Satterwaite 2007). Key industrial cities London and Berlin exemplify the rapidity of Eighteenth Century urban expansion: London's population grew from 960,000 in 1800 to 6.5 million in 1900 and Berlin from 172,000 to 2.7 million (Spielvogel 2012). City size has also increased greatly: the average population of the world's 100 largest cities in 1800 was around 185,000, in 1900 it was around 730,000 and by 2000 was 6.3 million (Satterwaite 2007). Urbanisation proceeded rapidly in the wake of industrialisation: in 1800, Britain was 40 per cent urban and rose to being 80 per cent urban by 1914 (Spielvogel 2012). Today, we talk of the vast urban conurbations of the mega-cities that extend over tens of thousands of square kilometres, with mega-cities across the world and a global population that is now predominantly urban. Modernisation of urban transport is also associated with rapid urban growth, both in history and in contemporary times. Although our popular image of the industrial city of the 1850s is one of squalor and overcrowding, there were comparatively few very large cities and these were mostly in Europe and Asia. In looking at the leading cities for urban transport development in Europe, the UK and the US, we need to remember that the cities of the nascent electric tramways, subways and railways have usually been swallowed by subsequent growth, incorporating what might have been relatively distant outlying towns and new suburbs from the major cities.

An insight from this condition, and one that is essential in understanding the history of urban transport, is that each transport mode has attributes that match different demands. Heavy rail, for instance, is suited to moving large numbers of passengers over longer distances at high speed, but this necessitates widely spaced stops. Shorter public transport trips are better served by light rail and buses but these modes cannot match the loading of heavy rail, and so forth. From the traveller's perspective, not all trips are the same as they vary by purpose, importance, duration and cost, and therefore engage a wide array of values, preferences and interests. Because there are so many different demands on urban transport systems, from both the operators' and users' perspectives, a range of technologies is required. This relationship is not just one-way; new technologies can generate new mobility options and opportunities, so that increasing mobility is not merely satisfying latent demands, but is also creating new demands for travel that did not exist, nor were predicted, prior to its introduction. For example, the bicycle boom of the 1890s saw a generation of urban women achieve a degree of independent mobility that was novel at that time, thereby becoming a social influence promoting women's emancipation (Furness 2010).

As a consequence of new transport technologies (and associated urban systems) and the growth in the types of, and opportunities for travel, there is an overall trend of increasing diversity of mobility and complexity in transport systems. Urban form and functions also become more complex as industrial cities evolve, a process promoted by transport, and one that also shapes travel patterns and demands. One of the ways in which transport creates new patterns and characteristics of demand is through its relationship with urban land use. For example, a new public transport stop can serve as the focus for a new set of journeys and opportunities, which might be as simple as that location attracting a new business with new employees and commercial logistics. Most travellers in the industrial cities of the 1850s had few mobility choices: they could walk, take a horse-drawn carriage or perhaps use a new horse-drawn omnibus and they may have had a railway service. Most passenger journeys would most likely be very local. By the start of the twentieth century, just two generations later, even urban travellers with relatively low incomes had an array of choices before them: railways, some subways, trams and buses of various kinds and the bicycle. People were travelling further, the costs of transport were lower and they were travelling for a wider range of purposes in cities that had grown outward, upward and in diversity and structure.

This historical overview proposes four phases, as shown in Table 1.1. All larger, developed nation cities have gone through these stages to some extent, as have most of the smaller cities and towns. Developing nation cities often present a more complicated picture, but usually also display features of the same general evolutionary themes of the mechanisation of mobility. Some cities have remained as 'transit cities' by retaining and enhancing public transport and not taking up the private car to a great extent, and there are many examples in Europe. But even for cities with strong public transport systems, it is likely that an older core has retained public transport and newer areas are dominated by roads and private vehicles. Older cities and older locations in cities may also have retained a premodern walking mobility system. We begin this account in the latter phase of the Industrial Revolution in the mid-1800s, that of the Victorian industrial city, for here many of the themes and influences that enabled subsequent motorisation of transport are put into place.

Table 1.1 An overview of modern urban transport systems

Phase of Emergent Technology	*Key Modes*	*Defining Features*	*Urban Identity*
Coal and Chaff: Semi-modern Mobility in the High Victorian Industrial City (1830–90)	Individual • Walking • Horse drawn Mass • Omnibuses and other horse-drawn vehicles • Steam railways	Railway boom (intra-city), first horsecars (trams)	• Industrial city • Palaeotechnic technologies • New railway towns
Spark and Spoke: Mechanised and Motorised Mass Mobility (1890–1920)	Individual • Walking • Bicycling Mass • Railways • Electric trams	New transport technologies, rise of mass mobility, first subways, first era of large transport infrastructure projects	• New suburbs
Petroleum-Powered Automobility: Private Transport Infrastructure and the Suburban City (1920–75)	Individual • Walking • Bicycling • Motor vehicles Mass • Railways • Electric trams • Trams • Buses	Rise of private car ownership, stasis and decline of public transport, growth of car-based urban forms	• Fordist
Automobility and the Post-Industrial City (1975 onwards)	Individual • Walking • Bicycling • Motor vehicles Mass • Railways • Electric trams • Trams • Buses	Flattening of car use in car-dominated cities, rise of service economies in OECD, revival of public transport	• Post-Fordist • New urbanism • Poly-centric city • Edge city • Post-industrial city

Coal and Chaff: Semi-Modern Mobility

In the classic era of the Victorian industrial city arising from the Industrial Revolution, urban and rural life is transformed, with rapid urban growth of established cities based on the new industries (iron, steel and textiles) and new cities centring on local industries and natural resources (coals, iron ore and water power). Urban transport in the leading economic nations throughout this period of rapid city expansion to around the 1870s lays the foundations for modern transport systems – but it remains essentially un-motorised. Most of the movement of goods and people in cities in the Victorian industrial city is by walking and horse-drawn vehicles; the only collective form of transport is the omnibus (i.e. horse-drawn carriages that begin operating in Europe and the US around 1820 (Post 2007)). City-dwellers walked a lot relative to their descendants in contemporary cities, and in cities reliant on walking, services are often clustered more evenly, so that walking provides convenient access to often-used services.

Omnibus services, comprising a carriage and one or more horses (or mules), began to ply their trade in US cities in the 1820s and persisted until around the 1880s and a little later in some places, generally finishing before the end of that century. Horse-drawn carriage technology evolved into horse-drawn omnibuses (known as 'horsecars') and began appearing in the 1830s and 1840s. Post (2007, p. 15): 'As would happen again and again with new modes of urban transit, many people thought that life was better because of the omnibus, and yet nobody was really satisfied.' Satisfaction frequently depended on the condition of the roads, as 'macadamising' using crushed stone was not yet universal, and in many settlements there were no public budgets for road paving. While steam railways were excluded from city streets, the technology of iron rails and flanged wheels did contribute to a major change in urban mobility by increasing the efficiency of horse-drawn (and mule-drawn) vehicles. Putting the horsecar on rails proved immediately popular as this resolved both the problem of uncomfortable rides on poor road surfaces and allowed greater loads to be pulled by horsepower. In 1832, a horsecar line from New York City to Harlem was opened. Also in the 1830s, lines opened in Philadelphia, Boston and Dresden and in the 1840s, in Berlin and Baltimore (Post 2007). Schatzberg (2001) states that by 1860 most US cities had networks of horse railroads that had rapidly replaced horse-drawn omnibuses. Two hundred cities had installed or had ordered streetcars in the US by 1890 (Nye 1997). Despite the technological changes that were evolving, the horsecar predominated until the last decade of the century; in 1890, 70 per cent of US urban rail transit was horse-drawn (Flink 1988).

Putting trams onto rails meant that trams could be pulled more efficiently, providing an incentive to make trams larger, so the first generation of horsecars were effectively railway carriages weighing up to four tons (Post 2007). When overloaded, these trams brought the unfortunate beasts of burden's lives to an early end, and spurred by the financial losses (with few concerns over animal cruelty), owners began to investigate building lighter trams 'to diminish the strain on the pocketbooks of the men that owned streetcars: A horse was a valuable commodity' (Post 2007, p. 18). Post (2007) states that by 1800 there were more than 3,000 miles of street railway with almost 20,000 horsecars in daily use in the US. These services proved very popular; Vuchic (2007) writes that in North America in 1882, there were 415 street railways companies, with 18,000 streetcars and 100,000 horses.

Initially, the new suburbs reached by horsecars were home to poorer families, at least in the US; subsequently, the suburbs became home to the middle classes of office workers and business people (Post 2007). As the distances from the city centres grew, the practical limits of horsecars were being reached. Conditions for the horses (and mules) drawing the horsecars were terrible and even with the lighter-weight generation of trams, average speeds were not high, being about four to five miles per hour. Paving provided for horsecars was usually cobblestones to aid traction, something unpopular with other vehicles (e.g. carriages and freight wagons). As Post (2007, p. 22) concluded: 'For all sorts of reasons, nobody thought of horse power as a long-term "solution" to the problem of mass transit.'

Motorisation of urban transport begins with the steam railways and the railway booms of the mid-to-late-1800s, but paradoxically, this occurs effectively in parallel with the pre-existing horse-powered modes and the reliance by most city-dwellers on walking for most journeys. It is paradoxical because although the Industrial Revolution is associated with the mechanisation of all manner of industrial, commercial and domestic processes and tasks, for the large part, passenger transport within cities is largely bypassed by mechanisation for the better part of the Industrial Revolution. Railways come into play for passenger mobility at the end of the Industrial Revolution in the UK and later in the nineteenth century in the US, Europe and a number of the European colonies and dominions.

For cities, the railway boom's initial significance was the speed and (relative) convenience of connections between cities compared with the horse-drawn stagecoach. Even relatively short stagecoach trips between cities could take several changes of horses and involve transfers between services. Point-to-point trips by rail, notwithstanding some of the discomforts that this initially involved, amounted to a revolution in mobility. Carriage services between cities and towns disappeared almost overnight as soon as competing railways arrived. Noted transport researcher, Vuchic, states (2007, p. 8):

> The benefits from railroads were so great that, following their introduction in the western countries around 1830 to 1840, construction of their networks proceeded rapidly; by the end of the nineteenth century, virtually all European and North American cities depended on railroad services for their economic functioning and growth.

Much was made at the time of the advantages of motive power over horsepower. Schivelbusch (1986, p. 7):

> Unlike traffic on the waterways, land traffic had until then been the weakest link in the chain of capitalist emancipation from the limits of organic nature, because animal power – on which land traffic was based – cannot be intensified above a fairly low level.

Up until the railway era, however, the organisation of the coach system in England had reached a high degree of organisation and sophistication and Schivelbusch (1986) reports that there were 342 scheduled daily departures in London in 1836 and the largest firm employed 2,000 and had 1,800 horses. As Schielbusch comments (1986, p. 7): 'These efforts did, in fact, introduce a trend that eventually made mechanization appear as the final logical step.'

As for timing, the rapidity of the growth of the steam railways in the mid-1800s was extraordinary, a boom in the UK known as 'railway mania'. In the UK, there were less than 100 miles of railway in 1830, but by 1852 the total had reached 6,600 miles and then further grew to over 14,000 by 1875 (Freeman 1999; Beaumont and Freeman 2007b). According to Freeman (1999), there were three main phases of UK railway investment: 1824–25, 1836–37 and 1845–47, with the last being the greatest. In 1830, the Liverpool–Manchester railway opened, being the world's first inter-city rail link. In the US, growth is similarly rapid but at a greater scale, with the bulk of the tracks laid in the latter half of the nineteenth century.

Railways were also a major factor in the social and cultural transformations wrought by the Industrial Revolution. Economically, the railways provided access to new markets and raw materials, fostering regional specialisation in manufacturing, increases in the scale of production and new industries and products, all having implications for urban development. New railways, although typically not inexpensive to use, provided new opportunities for travel for a wider range of the populace and for consumer goods, as Carter states (Carter, I. 2001, p. 9): 'Social space and social understandings were transformed, as train travel radically widened many people's physical horizons. Local production and marketing structures withered in the face of specialised rail-borne regional production tailored to national markets.' Contemporary and subsequent commentators, including Karl Marx and others, wrote of how the railways conquered space and time. Schivelbusch (1986, p. 12) suggested: 'As the sensory perception of exhaustion was lost, so was the perception of spatial distance.'

Aspects of the understanding of time and space were also brought into an industrial frame, such as by creating uniform time measurement to facilitate railway timetables across the railway systems. Jules Verne's 1873 novel *Around the World in 80 Days* was a panegyric to the possibilities of rapid long-distance travel by steam trains and steamships. New professions also arose in the service of railway design, construction and financing, and existing professions were re-oriented, evidenced through the new railway engineers and the creation of railway stocks. Manipulation of the landscape through engineering to accommodate the new railways, with bridges, tunnels, excavations and embankments aligned to engineering dictates of the new rail lines, was a development that distressed the Romantics of the era. Set against the loss of local traditions and desecration of the countryside were the highest of modernity's ideals, for the railroads, both symbolically and actually, were held to bring forth greater social equality and progress: for the Victorians, railroads were no less an enabler of civilisation.

Significantly, in Europe and the US, nearly all cities banned stream trains from urban streets. Few cities were exceptionally large at this time, so that locating railway stations away from city centres did not usually entail much of a journey to reach central locations. In some cities, such as New York and Philadelphia, the solution was to use horses to bring the carriages the final leg into the city centre (Schatzberg 2001). Inter-city rail travel was rarely along city streets. So passenger rail services of the period were oriented out of the cities and their impact on cities is in connecting those outside the city to the city, rather than facilitating mobility within cities. To contemporary minds, an unwillingness to have the smoke, stream, noise and not inconsiderable dangers of stream engines within cities is unremarkable; however, the cities of the time were already highly polluted, a state of affairs only worsened by the excrement of the large numbers of horses in service of urban mobility. Yet the railways were a major force in economic development and helped further boost the growth of these industrial cities, which in turn, increased the demand for horse-drawn vehicles to accommodate the growing populations and their mobility needs.

There was some urban railway development, however, and the first subways were constructed in this period, with their high levels of patronage proving the viability of mass transport within the city. Many early lines, such as London's Metropolitan railway that opened in 1863, were cut-and-cover projects running down existing streets (meaning a tunnel built by cutting an open trench and then building a cover on which the roadway was re-established), with tunnelling techniques developed as the century progressed. A variety of means were used to provide grade separation for urban rail, including elevated tracks, viaducts, open cuts and embankments. London's first line, using steam locomotives, carried 9.5 million passengers in its first year, 1863, but smoke from the locomotives made travel through tunnels unpleasant, to say the least. Other forms of motive power to steam locomotion were required and several types of electric locomotive were developed and put into use in the late 1800s, such as in Berlin, Boston, Budapest, New York and Paris. What is noteworthy about subway (and urban rail) construction is that it has continued worldwide since its inception, with large systems opening at regular intervals (e.g. the 1950s: Toronto 1954; the 1960s: Beijing 1969; the 1970s: Hong Kong 1979, Sao Paulo 1974, Washington DC 1976, San Francisco 1972; the 1980s: Singapore 1987; 1990s: Shanghai 1993; and the 2000s: Los Angeles 2000).

Although urban rail development has its genesis in this era, its diffusion occurs over quite a long time, even after the basic technologies are considered mature. Most subways are a product of the twentieth century, not the nineteenth, with prominent early-adopters in Berlin (1902), Boston (1896), Budapest (1896), Buenos Aires (1913), Istanbul (1875), New York (1904), Paris (1900) and Philadelphia (1907). In the decades between the world wars, many of the most famous systems were opened, such as Moscow (1935), Osaka (1933) and Tokyo (1927). Just as experimentation had resulted in the application of steam engines for motive power that promoted the growth of the railways, the invention of steel rails in the late 1850s was instrumental in the further growth of railway networks using larger locomotives and longer rail cars.

What emerged in these early decades of 1830–90 was a highly disaggregated arrangement of essentially local services and inter-city links, with a plethora of private owners and operators. However, such was the level of activity that all major and many smaller cities were interconnected by rail, with the larger cities having several competing firms and services. Yet, the historical trend of railway passengers is still relatively modest through the 1800s. Beaumont and Freeman (2007b, p. 15) stated:

> Most city termini proved unable to handle the rapidly expanding volumes of trains and traffic, goods as well as people, that railway operators sought to

> accommodate. The outcome was a wave of new terminal construction across European cities that saw the demolition of vast areas of central urban building, to be replaced by a series of grand terminal edifices, fed by swathes of approach lines and service yards.

Prior to 1870, there was the railway boom of the mid-century that saw an explosion in new lines and services between cities and urban centres and new railway towns established (usually at some distance from the cities), but these did not directly affect daily passenger trips in cities. Not much happens to urban transport technology within cities, despite the revolution brought by railways in the mid-1800s railway boom that provides a major economic impetus in industrial and economic growth and ends the era of canals and horse-drawn long-distance transport.

Mechanised and Motorised Mass Mobility in the Second Industrial Revolution

These sprawling, polluted, crowded and uncomfortable cities of the Industrial Revolution strained against the limits of their transport systems. By the time of the Second Industrial Revolution (1870–1914), the larger cities suffered from the effects of sustained and rapid economic and population growth. Streets of larger cities were congested with foot traffic, horse-drawn omnibuses and other carriages, a problem compounded by an absence of traffic controls. Nye (1997) reports that in the US in 1880 some 100,000 horses pulled 18,000 cars over 3,000 miles of track at an average speed of 5 mph, depositing 1 million pounds of manure. Horse use was expensive: tram owners owned many horses, which had to be fed and stabled, given veterinary care and looked after.

Cheape (1980, p. 2) writes of the US situation:

> In big cities as well as in large metropolises, the growth of population, area, and specialization heavily burdened public transit, that is facilities for transportation by common carrier in the municipality and its suburbs. A larger population meant more passengers. The number of rides per capita rose faster than population.

Several factors come together to compound the problem of population growth, such as the increasing specialisation in land use within the cities that tended to concentrate traffic in central business districts, while the rise of the suburbs required local transport to cover greater distances. Existing public transport technologies did not meet the demands being made for higher speeds, greater access to the suburbs and to carry greater numbers of passengers. Cheape's observations of the US also applied to the larger UK and European cities of the time (Cheape 1980, p. 2–3):

> The steam-powered commuter railroads, horse-drawn omnibuses, and horse railways developed for the mid-nineteenth century commercial city failed to

> meet these needs. Steam railroads serviced a small and decreasing fraction of local movement. Their massive engines were ill suited to frequent stops and starts. Antagonised by their noise and dirt and fearful of the dangers of explosion and collision, citizens fought to keep them off the streets and out of the downtown area. Steam railroads carried passengers from the outermost suburbs but could not offer transportation in and around the city. Commuter roads carried less than 10 percent of local traffic.

There is excessive competition for road space at the end of the nineteenth century, with the new modes of horsecars and older modes of horse-drawn carriages mixed with other horse-drawn vehicles and pedestrians; the early mass transit modes have only added to congestion and there is experimentation around the world to find an acceptable means to motorise mass transport.

As Cheape (1980) notes, in the last quarter of the nineteenth century, US public transport entrepreneurs could choose between three types of motive power in steam, cable and electricity and three forms of transport, being street railways, subways and elevated railways. Several US cities adopted cable cars during this period and by the mid-1890s there were some 625 miles of cable car lines nationwide (Flink 1988). Developing an effective design for the electric omnibus proved technically very challenging, with many competing designs in experimentation and use, such as using overhead wires, underground conduits and batteries on-board the trams. Noteworthy was the system built by Frank J. Sprague in Richmond, Virginia that began operating in 1887 and 1888. Sprague's model proved catalytic and prompted many other North American cities to take up electric street railways. Cheape (1980, p. 6) wrote: 'By 1890 electric car trackage was 15 percent of total trackage, more than double the cable figure. By 1902 it was 94 percent of total street railway mileage, and cable and horse lines accounted for only 1 percent each.'

Replacing horsecars with electric trams occurs through the decades of 1890–1930 in North America and Europe, with the introduction of electric trams across the cities of Asia and the Pacific, South America and Africa. Electrified trams in Europe lagged behind the US; by 1890, the US had 96 km of electrified lines (Vuchic 2007), but an investment boom in Europe followed suit, such as in Paris between 1890–1919 (Soppelsa 2009). In many places the change was rapid. In England, for example, the horse population peaked in 1901 at 3.25 million, a century after the beginning of the Industrial Revolution; a generation later, by 1924, this had fallen to less than 2 million (Garrison and Levison 2014). Tram design also continued to improve through this era, with vehicles becoming larger with more seats and wooden bodies replaced by steel. Lines are also extended into the suburbs with European cities adopting the tram as a primary means of urban transport. Electric tram lines replaced horsecar lines and were longer. In the US, electric trams were built by private investors, either by owners of horsecar lines or by speculating investors (particularly real estate investors).

Up to this time, urban transport is a derived demand and it exerts relatively little influence over other realms of life, on society or economic development.

As Flink (1988) noted of the US, in 1860, the 5 cent fare for horsecars was beyond the reach of the dollar-a-day workers and walking was the prevalent means of transport, with cities arranged around the limits of pedestrian locomotion. Beginning at this time, transport becomes also a major force in cities, shaping their form, function, and structure and exerting influences on social, economic and environmental variables.

It is the mechanisation of mobility that creates modern urban transport. For example, one of the advantages of the electric tram over the horse-drawn tram was its substantially higher speed and tramlines began to increase greatly in length. Urban development tended to follow these tramlines, often 20–30 km long, producing distinctive new urban forms. Urban function was altered as a result, with city centres' specialised functions being reinforced as they become more accessible to growing populations. Patterns of socioeconomic difference also become marked under the influence of transport, with the middle classes moving to the new suburbs and the working classes concentrating in the inner cities around industrial employment hubs. As Nye observes (1997, p. 137):

> The trolley had contradictory uses and meanings. Americans used it to escape from and to embrace the city, to standardize transportation for all classes and yet to establish differences in housing patterns. They both created and abandoned it in order to solve the problem of overcrowding. They first celebrated the ride as a novelty and later saw it as an unpleasant necessity. Their characteristic destinations—the workplace, the downtown, the countryside, and the amusement park—were environments with contradictory ethics: work, consumption, pastoral retreat, and escape into mechanized pleasures. On the streetcar, Americans journeyed into the fragmented worlds of modernity.

Bicycles are frequently omitted in urban transport history but during this era they play an important role in urban mobility and as agent of change. In some respects, it was the bicycle that displaced the horse, more so than the motor vehicle. As Furness (2010) states, the bicycle 'quickly became the standard by which horses were measured': compared to horses, it required no constant expenditure, no need for specialised care, and had all the advantages and none of the disadvantages of keeping a horse. City authorities had an incentive to replace horsing; in the 1880s, New York had to remove 18,000 horse carcases and 150,000 tons of horse manure annually (Furness 2010). After many decades of experimentation, the basic design of the modern bicycle was determined and combined with other factors, such as the invention of the pneumatic tyre, giving rise to a boom in bicycle sales in the 1890s.

Bicycles also highlighted a feature in the evolution of transport technologies; rather than just provide a different means of undertaking existing journeys, new technologies create new (and often unpredicted) opportunities for mobility (i.e. journeys that were not previously possible). As Sachs explained (1984, p. 104):

> The bicycle enlarged the immediate vicinity and multiplied the destinations that could be reached in a short period of time; whether to the factory to work or the lake for a swim, to church in the neighboring village or a flirtation in

> the woods, in the bicycle saddle one felt oneself the master of one's native territory. For the first time an achievement in transportation invigorated local life—for the railway had sooner enticed travellers to regard themselves as masters of the nation (and later, the airplane as masters of the world).

Bicycles had a prominent role in the women's emancipation movement, as it offered women independent mobility without a male chaperone; further, these freedoms were symbolically linked to emancipation and women cycling became a focus for debates over the behaviour and independence of women (Furness 2010). Bicycling allowed women to escape the physical and ideological confines of the home. Mechanising mobility greatly extended the range of personal transport, which had both social and economic implications. Furthermore, the bicycle democratised urban transport in several ways. It offered access across cities without need to hire horse-drawn vehicles or use of new electric trams; by the end of the 1890s, the costs of bicycles (which had been expensive) had fallen significantly (Furness 2010). Railways of the era, with their first-, second-, third-, and even fourth-class carriages, reinforced the era's class distinctions, whereas bicycles offered autonomous mobility without the class-based stigma of mass mobility.

Perhaps more contentiously in urban transport history, is the bicycle's role as a precursor to automobility, through its promotion of the idea of the potential for individualised long-distance transport, to its role in promoting road building, road law, repair centres and mass production. Sachs considers (1984, p. 106): 'The bicycle mobilized desire for an automobile.' But in doing so, the bicycle had sown the seeds of its own demise (Sachs 1984, p. 105):

> Because the bicycle did not move without muscle power, it remained an outsider among the new technologies with their aura of progress, for it was precisely in the substitution of mechanical power for muscular exertion that the point of progress was recognized at the time.

One reaction to the inhospitable city was flight, particularly by those who could afford re-location, and here transport was central. Basically, the suburbs of the Western world that grew between 1850–1920 were a product of urban rail, with those closer to city centres more likely to be connected through trolley cars and subways (Mumford 1961). Suburban development was often led by transport companies, electricity companies, as well as by land speculators. Oftentimes these railway suburbs were essentially new towns spaced according to railway stations and separated from cities by greenbelts of agricultural land. These new suburbs had a relatively concentrated form because services had to be accessible by walking.

Petroleum-Powered Automobility: A Story of the Rise of Private Transport and the Demise of Public Transport

Just as the railway was the iconic technology of the Victorian era, despite the fact that it had limited impact on travel within cities of the era, the motorcar indisputably is the technological symbol of the last century. Many have identified the twentieth century with the motorcar, giving rise to this era being identified as that of automobility. Automobility can have two meanings: first, self-propelled mobility, such as walking and cycling; second, 'the structures, institutions and policies which underpin and facilitate car driving' (Cahill 2010, p. 40). For many, automobility is shorthand for transport systems based on privately owned motorcars. For sociologists, automobility is also an ideology that entails the car and associations with freedom, progress, autonomy and notions of individual privacy (e.g. Bohm *et al.* 2006; Urry 2007). There are several points of inflexion on urban transport history during this era: the rise of the motorcar for mass mobility, the demise of tramways in many nations and the rise of buses, as well as the decline of active transport (walking and cycling).

Just as the railways demonstrated the limits of mobility using horses, the bicycle suffered a similar fate, being both motorised in the creation of the motorcycle and replaced when motorcars became cheaper and accessible to wider populations. As Sachs explains (1984, p. 105): 'There was nothing to be done about it, though—the bicycle fell victim to the contemporary view of what constituted technological progress: overcoming physical limitations through the power of the motor.' He finds the logic similar to that which found railways superior to horse-drawn carriages:

> Even though the bicycle multiplied bodily energy extremely efficiently and broadened people's arena of activity many times over, it remained captive to the defect of corporeality; over the long run, therefore, it would be at best a disagreeable substitute vehicle for the nonmotorized.
>
> (Sachs 1984, p. 106)

Initially, up until the beginning of WWI, the motorcar was a luxury item for the wealthy, frequently driven by chauffeurs, and its appearance on the roads and the dust from unsealed roads was often unpopular with other road users. Car ownership rates were highest in the world's wealthiest nations of the time, the US and then the UK. While there were many competing, small-scale motorcycle and motorcar manufacturers, the emerging commercial opportunity appreciated by a few was to capture a broader market. As Flink (1988, p. 33) observes:

> Recognizing that the upper-class market was nearing saturation and aware of a great demand for outmoded buggy-type cars and second-hand conventional vehicles, after 1905 the more enterprising American manufacturers, in contrast with their European counterparts, turned to the volume production of lower-priced cars for the developing middle-class market.

Henry Ford was the preeminent manufacturer in producing low-priced cars in high volumes. Mass production is also taken up in the UK, but cannot match the US in reaching mass markets (Flink 1988). Social change is also central to the growth of mass markets for motor vehicles, for while the First Industrial Revolution gave rise to the working class, the subsequent development of the middle classes in Western nations provided a social class with the financial means to purchase cars and motorcycles. During the inter-war period in the UK, car ownership becomes more widespread, but is only affordable to the middle classes (O'Connell 2014).

Motorcar ownership grew rapidly, first in the US and then the UK. At the turn of twentieth century, there were 8,000 cars in the US, but by 1905, this had grown to 79,000 and then to 10 million by 1921 (Ponting 1991) – it would take Europe until the 1950s to rival this level of ownership. Europe's boom in car ownership occurred between 1950–70. In 1904, there were less than 9,000 cars registered in the UK, rising to 132,000 in 1914, together with another 259,000 other public and private motor vehicles (O'Connell 2014). In this era of motorisation, there was high motorcycle ownership, higher than car ownership in the 1920s. As income levels lifted, UK car ownership rose from 100,000 in 1919 to almost 2 million in 1938 (O'Connell 2014). UK bus and coach-use also increased greatly during the inter-war period; between 1920–1938, the number of bus and coach passenger miles travelled increased from 3.5 million miles to 19 million (about the same as carried by the railways at this time) (O'Connell 2014). As with the railway era in the former times, there are many direct and indirect motor industry jobs in the UK, being almost 1.4 million jobs in 1939.

Faster cars caused more road collisions on roads that were considerably more mixed in road users and lawless than is the case today, so that in the 1920s it was reasonable to describe the roads as dangerous; fatality rates were higher than today in developed nations. Roads were increasingly becoming the domain of motor vehicles and the state moved to reduce the carnage of road trauma through various measures, such as driving tests and licences, speed limits and pedestrian crossings (all usually against opposition from motorists).

Private transport's gain was public transport's loss: car ownership and levels of use were inverse to public transport use. Naturally enough, this first occurred in the US; public transport ridership declined precipitously by 75 per cent by 1970 from a peak in 1908 (Yago 1984). For other developed nations, such a trend did not occur so dramatically until after WWII. In England local bus passenger journeys declined from 16 billion to 4 billion between 1950–2000 (Cahill 2010). Similar effects occurred in Australia (BITRE 2014) and elsewhere through the twentieth century, but the effects of public transport decline were not nearly as severe across Western Europe, even though car ownership rates increased substantially.

While suburban development was initially fostered by public transport, it greatly accelerated in scale and scope until the era when car ownership and use became common (Mumford 1961; Jackson 1985; Hall 1988). Unlike the suburban development created by the Victorian era railways outside the larger cities, where the railway stops served to concentrate urban functions and structures around specific nodes, road-based suburbanisation lacked this coherence and contributed to dispersed suburban settlement, with characteristic commercial development along roadways and the separation of residential, commercial and other land uses. This effect was revealed comparatively quickly when car ownership became more widespread; car-based suburbs were emerging in the UK and the US in the 1920s and 1930s.

WWI proved a difficult period for electric trams and in many ways portended their demise as a means of mass transport in much of the world, with a number of European cities being the exceptions. In the US and elsewhere, owners found themselves squeezed in a familiar vice where fares were fixed (often by city authorities/franchises) while operating costs were rising (such as for wages or fuels). It is no coincidence that in places with organised labour, pay disputes and industrial actions occurred, partly because operators had limited opportunities to pass through the costs of wage rises through fare rises and because drivers and conductors 'had to expect low wages and extended working hours and, usually, an adversarial relationship with customers' (Post 2007, p. 21). But motorisation proved more damaging; trams had relatively low operating costs, but buses did not incur the high costs of track repair and maintenance, and this factor is usually cited as the cause of the demise of trains in favour of busses. Between WWII and the 1960s, most of the trams in the US and the UK and many in Europe had been replaced with bus services (the last Paris tram ran in the 1930s and the last London trams in the early 1950s). It was not until the 1980s that there was a resurgence of trams in many cities, notably in the US (Cervero 1998). As Vuchic (2007) describes, in the US, there were some 35,000 buses operating in 1940, slightly less than the number of trams and 'rapid transit cars' and bus ridership was about one-half the ridership of these two other modes. By 1965, buses were the dominant street mode in most cities. Persistence of trams in several European cities were due to different circumstances, as Vuchic explained (2007, p. 24):

> In several other European countries, on the other hand, attitudes towards tramways were much more positive. The organisational and financial situations of transit agencies were more stable, since transit systems were usually consolidated into single, municipally owned agencies. Separate tramway rights-of-way in many German, Dutch, Swiss, Austrian and other central European cities were preserved, upgraded, and in many cases extended.

Motoring and car ownership declined during WWII in Europe and North America, particularly as many nations had petrol rationing and automobile manufacturing re-directed into war industries. But in the post-WWII years of the 'long boom' in Western economic growth and national prosperity, private car ownership soared. Town planners generally did not anticipate the surge of car ownership following WWII, when car ownership expanded across the socioeconomic divides and mass ownership was underway. This was the era of the suburb in developed nations, especially so in the New World nations that experienced high rates of urban population growth; many have been described as 'suburban nations' (e.g. Duany and Plater-Zyberk 2001). Suburbs of this era differed markedly from their predecessors served by public transport. Public transport suburbs pre-WWII usually contained industries and factories, often with concentrations of different ethnic and cultural groups, included working-class suburbs, and had both planned and unplanned developments. Pre-war car-based suburbs there were also more modest and unplanned developments. Post-war car-based suburbs tended to be more socially and economically homogenous; they were without local industry or significant employment, necessitating car-based commuting. Despite the popularity of the new suburbs with the middle-classes, some critics were questioning and rejecting this urban form (such as by US critics Lewis Mumford and Jane Jacobs, albeit with contrasting alternative visions for ideal urban form), a movement that was only to grow in the subsequent era.

When the long boom came to an end, so too did the era of unfettered belief in the primacy of the private car as the ultimate provider of urban mobility. Significantly, the end of both of these much-lauded eras shared a common causal factor, namely the oil crises of the 1970s. What changed was the belief that privately-owned vehicles with internal combustion engines using energy from fossil fuels was the final stage of urban transport. Since that time, a succession of changes and developments has reinforced this change: climate change, peak oil, equity in access to mobility, competition for urban space, road gridlock and road trauma. We still live in cities in the developed world with high rates of private car use and the leading nations of the developing world are following the car ownership trend as their national incomes rise. However, few now seriously suggest that future car ownership rates in any major nation will rival those of the US or that mass mobility in future cities is best served by systems based on automobility.

Post-Automobile Automobility and the Post-Industrial City

Arguably, the era of car automobility has ended despite the continued worldwide growth in vehicle fleets, particularly in the rapidly industrialising nations of China and India. There are a couple of versions of the 'end of the car' thesis. Essentially, what has ended is the viability of both the rationale of the car for providing mass transport needs in city and the unquestioning acceptance of the virtues that have made individual car travel so attractive. Environmentally sustainable transport has emerged as the alternative to car automobility and it is the environmental critique that has profoundly concluded the automobile era (an issue pursued in greater detail in Chapter 4). In this section, we examine some of the other reasons why automobile automobility has ended.

Newman and Kenworthy (2000), Kay (1997) and others have argued that many of the claims made for the inevitability of car-based automobility are mythical. Central in this debate is the association between rising incomes and car ownership, so strong that rising incomes inevitably result in higher car ownership. They reject this claim on the basis that in many developed nations, the most expensive areas for home ownership have lower car ownership rates than less wealthy districts. Other associations with automobility are also dismissed on empirical grounds, such as the positive effects of warmer climates and suburban developments, that more spacious nations have higher motor vehicle dependence, and that land speculators create automobility and cannot be stopped.

As described above, a key attraction of car use is speed and that all other factors being equal, faster travel is always to be preferred. This may or may not be true as far as it goes, but all factors are not equal when we compare travel between competing modes and this prompted inquiry into various aspects of time and car use. In the 1970s, social theorist Ivan Illich (1974) raised a critical point by widening the conceptual understanding of the investment in time that car travel requiring by pointing to not only the time spent in cars when moving and when standing still, but also in accessing the car, and then in working to raise the capital to pay for owning and operating the vehicle; he said the average American 'devoted' over 1600 hours annually to the car. In a famous passage, he wrote (Illich 1974, p. 31):

> What distinguishes the traffic in rich countries from traffic in poor countries is not more mileage per hour of lifetime for the majority, but more hours of compulsory consumption of high doses of energy, packaged and unequally distributed by the transportation industry.

Taking these factors into a fuller account of vehicle speed, the car does not offer particularly high-speed travel; Whitelegg (1993) found that over a 20-kilometre trip, the 'social speed' (aka effective speed) of the car traveller required 70 minutes compared to that of a cyclist of 32 minutes.

Underlying the attraction of higher travel speeds is the liberation of saved travel time over slower trips. In practice, however, such savings have proved highly elusive and this has further weakened the case for car travel offering time savings over other transport modes and over other ways making connections and exchanges. Taking transport as a derived demand, the purpose of transport is that mobility provides access to places that we desire to reach and travel time is a factor that we take into account along with many others concerning how we undertake any trip; in other words, higher travel speed is not the primary objective in travelling. Although time budgets have been somewhat neglected in transport research, it is clear as a generalisation that as urban travel speeds have become quicker, people travel further but do not necessarily make more trips. Household trips for commuting, shopping, commercial purposes, social and recreational purposes, for example, tend not to increase when there is access to faster transport. Critically, the trend under car automobility (and this probably also happened in the era when suburban railways arose) is that a trade-off occurs; travellers exchange time savings from faster travel into further travel. This effect is tied to both increasing scale in cities and urban planning that produces greater dispersal of potential travel destinations, and clearly there is a circular logic operating here, with car transport encouraging longer trips and cities being designed on the presumption of car travel. Greater investment in road infrastructure does not lead to overall reduced time spent travelling by car drivers and passengers within cities.

Continued growth of the car-based suburb was tied to the critique of car-based urban transport systems as this urban form is predicated on access to motor vehicles. For many, the untrammelled expansion of low-density, free-standing housing of the middle classes on the ever-expanding urban fringe (and Greenfield developments adjacent to cities) was a social, economic and environmental failure that became identified as 'urban sprawl'. Critiques of car-based suburban development became especially numerous and more pointed since the 1980s, especially within the US (e.g. Jackson 1985, Fishman 1987, Rome 2001).

Criticisms include the large 'ecological footprint' of suburbs, loss of agricultural lands and natural habitats, promotion of car-dependency, high costs of development, implications for health, low aesthetic values and low cultural development. Of these, car dependency – the condition where transport and land use favours car use and provides few other mobility options so that a reliance on cars in unavoidable – is of particular interest. Absolute car dependency in transport systems does not exist, but many locations are highly dependent on cars for mobility. Car dependency is associated with high levels of car ownership and use (including number and length of total trips), very poor public and active transport options, dispersed land use, transport system facilities and services predominated by road investments, road design to optimise motor vehicle use and various subsidies and incentives to encourage car use (Newman and Kenworthy 1999, 2000; Jones 2011). A great many environmental and social costs are created by this condition that Soron (2009) called a type of 'compulsory consumption'. In those places where these transport costs are highest, there is invariably a high level of car dependency. Of particular concern are the mobility needs of those who have no access to motor cars for reasons of low income, age, physical or mental handicap and other factors. Car dependency creates mobility disadvantages and exacerbates existing inequities between those who are able to enjoy car use (with its high levels of mobility and public subsidies) and those without such access in transport systems with poorer levels of public transport services. Older debates also re-emerged from these concerns, noticeably that of suburban growth promoting increased inequity within cities by causing increased geographical concentrations of disadvantaged citizens (e.g. Squires 2002).

Smart growth, New Urbanism and other urban planning and design responses have sought to re-direct urban development in ways to avoid the problems of car-based suburbs through strategies of higher residential densities, mixed land uses and greater pedestrian accessibility, based on ideas put forward by Calthorpe (1993), Duany and Plater-Zyberk (2001) and others. While some of these concepts feature higher use of public and active transport (e.g. Transit-Oriented-Development), many retain the private motor vehicle as the primary means of transport. One striking lesson from the ongoing debate over these issues is the importance of the institutional divide between urban planning and public transport planning; progress in providing public transport access to the suburbs depends greatly on bridging this gap.

Often lost in the developed world's embrace of private car use was any questioning as to how fair the outcomes of mass car ownership were. For those without access to private transport, there was public transport as the mobility of last resort; i.e. public transport's role was to provide mobility social welfare. Not that such a sentiment made its way into public policy texts, but given the skewed character of transport investment towards road infrastructure and the neglect of public transport, the message was clear enough. In the 1960s and more strongly thereafter, social scientists and social welfare advocates began to investigate access to transport and found that societies with a high dependency on cars for mobility had highly unequal access to mobility. A condition of car-dependency had emerged, whereby households could only achieve high levels of accessibility through car ownership; in the post-war car-based suburbs, public transport was absent, widely dispersed, of low frequency, localised or insufficient in other ways to meet household travel needs. Within the wealthy developed nations, those places with high levels of car use exhibited strong discrimination against those unable to enjoy access to private cars. Politically, such discrimination was underpinned by investments of public funds into road infrastructure, funds that could have been directed towards public transport that would have reduced the gaps in equitable access to urban mobility.

Another of the aspects of the decline of support for urban freeway construction has been the realisation that public funding (i.e. state-funded) support for freeways was one of the greatest examples of middle- and upper-class welfare. Once the access-and-equity test had been applied to urban transport, governments could no longer axiomatically assume that increasing the convenience of car drivers was a net social benefit. Car-dependency did not only discriminate against the socially and economically disadvantaged, but state investments in road infrastructure and specific land planning decisions by state agencies were creating and increasing the burdens of disadvantaged groups. Transport disadvantage was manifested not only as a function of geographical location, but also of social structure (as by income, race, gender, age and other factors). Whatever automobile automobility promised, it clearly did not provide an answer to the growing concerns over access and equity, and this realisation marked a shift in transport discourse that placed access in a more central position.

Growth of the area covered by cities is obviously greater when settlement is dispersed under the suburban model of car-based suburbs than when concentrated. Rather than entering into the fraught urban debate over residential density, we wish to raise a different point over the implications of car-based transport systems in cities and urban space. Higher-speed transport requires more space than slower-speed transport, so that the greater the use of high-speed transport in cities, the greater the loss of space to transport infrastructure. In the era of automobile automobility, the intrusion of freeways into the existing urban fabric was rationalised as the inevitable price of progress, with high-speed vehicle travel given priority over the associated costs, such as lost open space, lost residential housing, loss of urban amenity and increases in air and noise pollution. Urban freeways serve as impermeable barriers in the streetscape (producing 'severance'), so the price of increased cross-city movement is reduced local mobility. Nearly all the major cities of the developed world, and similarly in the developing world, have been penetrated by freeways and major roads with great loss of urban amenity. Freeways can go underground, but the costs are high and can be very high; Boston's 'big dig' to accommodate a few kilometres of interstate and local freeways ended up costing around US$14.8 billion amid great controversy (e.g. Brown-West 2007).

There are few celebrations of such freeway intrusions now and there is something of a worldwide movement for freeway removal. Seoul's removal of the major elevated freeway over the Cheonggyecheon River and subsequent restoration into an urban park, opened in 2005, has become a celebrated piece of urban design worldwide (Schiller *et al.* 2010). Less well publicised but also important were the transport aspects of the initiative that involved the building of the Rapid Bus Transit system in Seoul and upgrading the regular bus system to compensate for the loss of motor vehicle access. Citizen opposition halted construction of the San Francisco Embarcadero Freeway in the 1960s and following later earthquake damage, it was torn down and replaced with a smaller roadway, public space and mixed-use development, to wide acclaim. A part of the reasoning for many proposals for urban freeway removal is economic; many of these structures are approaching their end of life spans and the costs of replacement concern local and state authorities. Urban freeways never really enjoyed universal support, but governments were generally happy enough to push the ribbons of concrete and asphalt through the cities until the 1960s, when a number of community protests succeeded in preventing freeway construction. Carlson (1995), for example, describes a number of case studies where community action prevented or reduced the impacts of a variety of US transport projects. Freeway building is far from over around the world, but in the developed world, presumptions that freeways have a net social benefit no longer hold (Newman and Kenworthy 1999). Governments and developers have to work harder to convince sceptical electorates that the benefits to motorists arising from new freeways trump the interests of communities and taxpayers that will bear the costs.

Developing world cities are particularly afflicted with transport problems. Greater private motorisation is choking the streets of the larger cities, with congestion levels and problems that can only be properly comprehended through personal experience. Cities such as Bangkok, Buenos Aires, Cairo, Jakarta, Mumbai and Shanghai are but a few of the large urban conurbations where there are many signs of economic and social development, but where road transport conditions are either not improving or are worsening (Dimitirou and Gakenheimer 2011; UN-Habitat 2013; Vasconcellos 2013).

Further to the diminishing of the rationale for basing urban transport on automobile automobility are signs that car use is falling in those places where car ownership is highest, namely North America, Europe and other OECD nations. At, or around the mid-2000s, the per capita mileage across the US, European nations and Australia peaked and began declining (Schipper 2011; Goodwin 2012). In other words, up to this time, there were annual increases in personal driving, but now there is slightly less individual driving each year; in nations with stagnant populations, total vehicle kilometres travelled is falling (but where population growth is sufficiently high, total vehicle kilometres travelled continues to increase). This phenomenon of saturated car use has recently become known as 'peak car', as David Metz (Metz 2014) describes it.

Experts disagree over the causes of the decline in driving in western nations and the available evidence is not yet conclusive. In all likelihood, the explanation is probably based on social trends and several of the suggested causes are contributing jointly. It does appear that young adults are not acquiring drivers' licences and owning cars at the same rate as previous generations. Another causal suggestion is that younger people's incomes have been reduced following the global financial crisis and they cannot afford car ownership and use. Delayed family formation amongst the younger generation may also be influential. Another explanation is that the inner-city living has become fashionable with younger generations and car ownership is unnecessary in such public-transport rich locations. Other sorts of economic changes may be changing transport habits, such as the influence of the Internet. What we might also be witnessing is a natural saturation effect, that we have simply reached the limit of the time that citizens in developed nations are prepared to spend driving (and car travel is avoided by choosing closer destinations, increasing trip chaining or exercising individual forms of travel demand management) (Metz 2008).

In sum, many of the beliefs in automobile automobility that made individual motorised transport so attractive during the golden age of motoring simply no longer hold up. Motoring was sold as a form of freedom and in its ideal condition, motoring does indeed convey a form of freedom in mobility. Automobile automobility in practice, however, is in many ways the antithesis of freedom, taking away mobility choices for many, restricting some to a condition of car dependency and shifting the costs of motoring inequitably within society and onto the environment and future generations, thereby eroding their choices and opportunities.

Tracks and Roads to Modernity

As suggested above, the overarching frame of modern urban transport's evolution has been that of modernity, although in contemporary times, some scholars have raised the idea that modernity is over and society has entered a 'post-modern phase'. For the most part, while modern transport histories record the transition of technologies and the circumstances surrounding these changes, overarching explanations are rarely undertaken. In the nineteenth century, there was little trouble explaining such changes, as the doctrine of progress was so strong and all technological development could be fitted into its narrative. Largely this technological progression was one of increasing mechanisation and increasing vehicle performance with efficient harvesting of energy sources, together with advancing engineering to create the infrastructure and control systems needed to assemble complex transport systems. It follows that the types and sources of energy used in transport tended towards those of increasingly lower market cost, of greater portability and of increasingly concentrated form (i.e. higher energy 'densities'). Harnessing energy in new ways allowed many of the transport attributes of progress to be realised, such as moving vehicles faster and covering greater distances, with lower costs and higher efficiency of energy use, of carrying greater numbers of passengers and of higher acceleration that allows for more stopping and starting.

Progress, in the nineteenth century, meant a great deal more than obvious improvements in selected metrics of passenger, vehicle and network performance, as progress was a motif of the First and Second Industrial Revolutions. Progress was a value of great symbolic significance, unifying ideas of improvements in social life through scientific, technological and social reforms. Railways were a potent symbol of progress in this era because of the rapidity of the changes they brought through such effects as closing the gap between cities and regions and between cities, in promoting new businesses and trades and by boosting material prosperity.

Although the railways have remained an integral part of modern transport, the widespread belief in progress – at least in its Victorian form – has not persisted. Of course, the idea of progress is an ancient one, but the Victorians saw their place as the development of civilisation as a consequence of the Industrial Revolution. That technology could be axiomatically taken as a social good seems extraordinary in contemporary times, considering the social and environmental costs of the industrial revolutions. However, a modern audience might find a parallel in the contemporary embrace of modern communication and information technologies largely unconcerned with exploring its negative consequences. During those earlier times, progresses' costs were either ignored or deemed resolvable through 'scientific' approaches, such as evidenced by the hygiene movement (such as urban water supply and wastewater and sewage treatment systems). Although progress still has many adherents today, the contradictions of technology overcame, in a sense, the belief that progress was universal, continual, assured and beneficial. WWI marked the first mass mechanisation of warfare and the harnessing of new technologies, notably from chemistry, was to many intellectuals, the end of progress.

What ended with this view of progress was the rhetoric of progress, with its promotion as a secular faith and its idealising cast. Although progress might have been revealed as a myth and the language of progress altered in light of its failings and contradictions, the underlying conditions of industrial societies were unaltered. Nor were only the fundamental aspects of modern society undisturbed; the directions that industrial societies adopted around the world shared many basic social, technological and environmental attributes and newly industrialising nations have taken up the same path, that of modernisation. Politically, institutionally and culturally, what unifies the practices under modernisation is the idea of modernity.

Studies of different modes of modern urban transport have made the link to modernity: railroads (Beaumont and Freeman 2007a; Schivelbusch 1986), automobiles (Sachs 1984; Ross 1996; Wolfe 2010) and bicycles (Norcliffe 2001). Urban transport has been, and continues to be, a force for modernity. It is also something that is shaped by modernity as it responds to the cultural, social, economic and other influences in the contemporary world. Modern transport technologies promote and express several key themes of modernity, including its contributions to urbanisation, capitalism, individualism, perceptions of time and space, as well as its lasting influences on industrial production (through Fordism), corporate organisation, and public/private relationships and the role of the state in fostering private enterprise, infrastructure investment and other factors. There are also arguments made that mechanised mobility itself is a feature of modernity (e.g. Urry 2007).

Economic histories of the last two centuries have recorded the influence of transport technologies in fostering economic growth and development, particularly through the railways and motor vehicles, so the associations between transport, urbanisation and the development of capitalism are well established, particularly in studies of the UK, US and Europe (notably of France and Germany). As we know, however, mobility in many cities seriously declined as a result of rapid growth during the nineteenth century, so that while economic growth in the form of increases in employment, productivity and outputs were sought by cities and states, there were accompanying social and economic costs, a portion of which were created by the transport sector, such as crowding, accidents and pollution. A reckoning of transport's material contributions to urbanisation and capitalistic growth came also and included the costs of these activities, such as urban squalor and overcrowding, ill health, urban pollution and workforce exploitation.

Speed has also been a hallmark of modern transport and a constant rationale for its evolution; higher speeds are held to be always desirable over slower speeds for all travellers. Speed, as a goal, is largely a modern phenomenon in land travel, as Urry (2007, p. 99) observed: 'Before the railway and its mechanization of movement, speed was mostly not a great issue.' Like all such tropes, exceptions and contradictions are readily found. For example, higher vehicle speeds are not desirable in the cases of collisions; further, the risk of collisions increases with vehicle speeds. Equally, fuel (and therefore energy) use and pollutant emissions increase with vehicle speeds, as does the cost of vehicle operation. Higher speed vehicles also require more space for operating, with the result that infrastructure for higher speed travel is more expensive, consumes more urban space and is more physically disruptive than lower speed infrastructure. For Illich, speed in transport systems is a problem in itself (Illich 1974, p. 24): 'High speed is the critical factor which makes transportation socially destructive.'

Modernity is also deemed to have influenced human consciousness in a variety of ways; the modern world is inhabited by modern societies. Interestingly, one of the key characteristics of modernity featuring in contemporary accounts was widely suggested in the first era of the railways, such as by Karl Marx amongst others, namely the influence of rapid transport on the perceptions of time and space (Schivelbusch 1986). Conceptions of distance and travel time for land transport were enormously contracted for the first generations experiencing such mechanised transport, changing perceptions long shaped by walking, horse-riding and horse-drawn vehicles.

Travelling at high speeds in mechanised vehicles is exhilarating and euphoric, but such feelings are mixed with other emotions and reactions (Sachs 1984; Schivelbusch 1986). Such sensations were often disorientating and unsettling, but train travel also liberated the traveller in other respects, by providing for leisure time while in transit, further making such journeys separate from normal reality. Adding to such feelings was the transformation of the traveller into an inert being – mobility occurring without input from the traveller – but also a dependent being, totally reliant on a technology that required no input from the traveller. Critic John Ruskin (in Chapter 4 of his *The Seven Lamps of Architecture*) was one of many at the time who wrote of the dehumanising effects of railway travel (remembering that steam train travel was typically very noisy and jarring). Such reports of alienation extend to the understanding of the passing landscapes viewed by passengers, whilst others in the Romantic movement described the linear routes of railway line alignments as intrusive and inimical to the natural contours of the countryside.

Similar reasoning has been advanced over motor vehicle driving. Ross (1996, p. 20), for example, writes: 'The various practices associated with driving cars are similarly "outside time". The post-war period of France's motorisation is also the moment when what Henri Lefebvre calls "constrained time" increases dramatically.' Mechanised transport is taken as the industrialisation of mobility in which travellers are but another component and this identity is imprinted on social consciousness.

Individualism is often taken as a hallmark of modernity. Sachs (1984, p. 99) identifies a trend in technology, a 'design gradient favoring individualization runs through the whole history of modern technology', of which the transition from railways to automobiles is a spectacular example. He describes how the automobile became the focus in a general social trend towards individualisation. As mass transit grew, so too did the impulse for individual travel and the motorcar enabled such preferences (Sachs 1984, p. 98): 'The characteristics of the automobile confirm the idea of independence and allow it to appear as natural.' Contrasting with the railways, the car could be acquired as private property, be at the owner's disposal, could travel to destinations of the owner's choice, need not be shared, offered quick travel and conveyed status on owners. As Sachs (1984, p. 99) describes it:

> Equipped with this dowry of design, the automobile is fit for marriage with the desire for an unfettered lifestyle, a desire that can be undermined only by the experience of one person's craving for freedom colliding with that of another, resulting in traffic jams everywhere.

A key expression of individualism expressed in modern transport has been the evocation of freedom. Freedom has been a psychological and symbolic motive force in automobility, reaching its apotheosis in the motorcar (Urry 2007, Seiler 2009). Through distance, mobility offers the freedom of liberation to escape the stasis of the local and the known, to find distant pleasures and to realise latent expectations. Yet the open-ended promise of freedom of mobility offered by motor vehicle use is illusionary. For a start, such a freedom is deeply contradictory, for the closer to universal freedom of movement one gets in a society where all enjoy the same freedom, the allure of freedom is diminished; you are going to places that others have left. Some aspects of mobility seem associated with reduced freedom; with greater mobility, the distinctive character of local places is reduced (the globalisation effect) and the novelty of journeys is lost through repetition. In daily life, mobility needs are not shaped by fantasy opportunities to travel anywhere at will, but by more mundane and routine trips determined by more obligatory requirements rather than those of leisure. Most of these daily trips, from commuting and shopping to education and personal services (such as trips to banks, post offices or clinics) do not seem particularly discretionary. There are many accounts of the thrill of independent travel by those driving for the first time, but this usually coincides with the period of young adulthood that is often the most unencumbered time of adult life. It is usually but a few years into adulthood where mobility is less about what could be done and more about what needs to be done. Lastly, open-ended choice can also be disarming; psychologist Schwartz (2004) identifies a 'paradox of choice' arising from an excess of choices requiring more decision-making and the resulting production of stress, dissatisfaction and depression (an idea with roots in Herbert Simon's maximisers and satisficers concept of the 1950s), although Schwartz's thesis remains controversial.

Where individuals and households are car dependent, motorcar ownership, as suggested above, diminishes freedom of mobility by reducing choices and opportunities for travel. Deprived of public transport and close destinations, mobility needs make car owners a slave to the vehicle. Autonomy in mobility is certainly attractive, but, as we know, there are many aspects to mobility and autonomy of individual automobility may be easily outweighed by collective mobility offering lower costs, greater comfort, more convenience and lower total travel time. Freedom may still feature in the marketing of motorcars, but this seems anachronistic in an era when mass motorcar use entails experiences that seem far from liberating and exhilarating, such as crawling along in heavy traffic, losing time driving that could be used for other activities and spending time and money having the vehicle serviced, cleaned and maintained.

As described above, motor vehicles are part of an elaborate technological system, and this system contributed greatly to two key institutions in forming and maintaining modernity, namely the organisation of manufacturing and the engagement of the state in enabling large-scale infrastructure. Henry Ford's production line for the Ford motorcar became the template for organising industrial manufacturing on a mass scale, i.e. Fordism. As David Harvey and others have explained, the importance of Ford's insights were not the assembly line production and other efficiency initiatives:

> [w]hat was special about Ford (and what ultimately separates Fordism from Taylorism), was his vision, his explicit recognition that mass production meant mass consumption, a new system of labour power, a new politics of labour control and management, a new aesthetics and psychology, in short, a new kind of rationalized, modernist, and populist democratic society.
>
> (Harvey 1989, p. 125–126)

In describing the French car industry, Ross (1996) states that it was the most vital in the nation's post-WWII economic growth and that increasing buying power of the era was used for car purchases:

> But to recount these simple factors is already to argue that the car *is* the commodity form as such in the twentieth century, an argument that becomes all the more convincing when we remember that 'Taylorization'—the assembly

> line, vertical integration of production, the interchangeability of workers, the standardisation of tools and materials—'Taylorization' was developed *in the process of producing* the 'car for the masses' and not the inverse.
>
> (Ross 1996, p. 19)

Modernity has a dual character: it unifies and standardises but also produces divisive and often unequal social effects, has negative effects that offset its benefits and has proved to be deeply ambivalent and contradictory. Railways and automobiles have both been lauded as symbols and agents of progress and civilisation during their era: railways in the high Industrial Revolution and cars through the era of mass ownership in the twentieth century. Both technologies were held to carry forward the project of modernity. Both technologies too, also came to represent the failings of modernity and be regarded as agents of the failures and contested character of progress. For example:

> For all that it appears to be inseparable from the pervasive ideology of rational progress in the nineteenth and early twentieth centuries, the railway has throughout its history generated strikingly incongruous effects on human consciousness. … So if the railway has shaped the ideology of progress, it has also shaped what Susan Sontag has (in a slightly different context) referred to as 'the imagination of disaster'.
>
> (Beaumont and Freeman 2007b, p. 13)

There arises the issue of whether there are other overarching directions in the evolution of urban transport. In contemporary times, four expectations shape the progressive agenda for urban transport, reflecting in response social justice and environmental justice: 1) Increasing social equity in urban mobility, 2) Reducing death and injury from transport, 3) Reducing energy consumption and using energy sources that are secure and renewable, and 4) Reducing environmental impacts.

2 Transport Governance and Institutions

As the history of urban transport shows, there are strong connections between the technologies of motorised mobility, the modes of urban transport, land use planning and city form and function. Opportunities and problems arose from these historical developments to which national, state, and city governments responded. Of particular interest in these responses are the governance and public policy institutions. This chapter examines some of the reasons that motivated government intervention in urban transport and gave rise to the common models of ownership that feature public and private ownership.

Times of Change

Unregulated markets in urban transport did not last very long in the early phase of the modern transport era; anywhere with the capacity for bringing transport companies under government control did so quickly after motorisation and mechanisation began. Creating, operating and managing large urban transport systems effectively was as much a matter of institutional design as it was of mastering the new technologies. This same lesson still applies to developing nation cities in contemporary times who have found that failings in governance lie at the root of many of their transport problems, more so than technical and economic factors (Dimitirou and Gakenheimer 2011). Historically, the rise and expanding reach of governments' bureaucratic power, governments' interest in overcoming and circumventing corporate power and the problems of uncontrolled transport development were often closely associated. England was a railway pioneer and where the early efforts at controlling the industry emerged, set a pattern repeated around the world in different ways. As Carter states:

> Appalled by fall-out from the mid-1840s Railway Mania, that unregulated speculative bubble, between 1855 and 1862 the British Parliament legislated to perfect the limited liability joint stock company, calming investors' qualms while satisfying railway companies' vast hunger for capital to fund construction and operation. This sturdy envelope has stood time's test: bureaucratised capitalism in its (to date) fullest flowering, today's giant limited liability railway company.
>
> (Carter, I. 2001, p. 10)

With the invention of the fixed rails for city trams, city authorities around the world granted monopoly rights to their corporate owners, but also moved to enact restrictions and conditions on these businesses, such as the control over fares. Once established, the relations between governments and corporations were generally very stable (which take quite a variety of different forms), with usually only crises, such as corporate failure, necessitating any major changes to existing relationships. Governments' activities in owning, running and controlling urban passenger systems was the norm and attracted little interest from scholars of government and public policy advocates (local controversies aside), until comparatively recent times.

Prior to these developments, the major preoccupation of transport authorities and their political leaders centred on operational concerns in public transport and the road system, together with regulating commercial and safety matters. There had always been, however, a considerable and complex set of governance institutions guiding and controlling transport policy and planning, research and development, transport law, financial management and relationships with other government agencies. Through the history of urban transport, state agencies amassed political and financial power, becoming the repositories of considerable expertise and experience. Over a century or so, the operational entities of the state and particularly those dealing with public transport, had achieved a high degree of vertical integration, notably when these took the form of state-owned enterprises. Many were involved in designing and constructing rolling stock and infrastructure; managing numerous businesses associated with railway stations and termini; had marketing, advertising and public relations departments; cleaned and maintained rolling stock and buildings; and were engaged in research and development and in their own education and training programmes. In city and state governments, the transport portfolio was typical amongst the most important in terms of expenditures, public sector employment and influence on the wider economy.

Additional to these political changes were other growing and changing demands on those responsible for governing urban transport. Urban growth, greater concerns with social welfare, environmental protection, new transport technologies and more varied transport demands increased the expectations of urban transport. This occurs in more complex cities where the performance of transport institutions is mediated by other urban systems, a problem that evokes political and governance challenges. While operational matters may be largely confined within transport agencies, other matters, such as infrastructure provision and land use planning, necessarily require coordination and integration with other agencies and stakeholders. Increasingly, it became apparent that effective institutions often had a comprehensive span of responsibilities (including management, planning, investment, operations and regulation) that had a great bearing on the performance of urban transport. These changes brought into question fundamental issues around the role and purpose of government in urban transport, matters that had been dormant for around a century.

Two developments in the recent past greatly increased interest in how urban transport systems are governed, the first of which was the political movement of neoliberalism and its associated programme of privatisation, particularly in the English-speaking developed nations in the 1990s. Neoliberalism challenged the model of centralised governmental control of urban transport through state agencies and state-managed corporations, bringing forward a re-reckoning of the divide between the state and corporate spheres. In those places where there was already a stronger involvement of private enterprise, such as in many Asian cities and nations, there was less interest in such policy debates.

This political development fitted with the second development, which was the culmination of many other processes and issues that gave rise to changing financial circumstances of cities and states. Prompted by successive city, national and financial crises (such as that of the global recession of the late 1980s and early 1990s) and widespread city indebtedness in developed nations, there was greater interest in how governments could reduce the costs of transport. In many nations, public transport usage was at the end of a long period of post-war decline associated with the rise of the motorcar and the associated vicious cycle of declining revenues, increasing public subsidies and cost-cutting that reduced services that in turn reduced patronage, causing further financial losses. Public authorities were also increasingly concerned with the rising costs of road infrastructure and its associated systems. These factors fostered innovation in public sector financing and new ways for attracting investments from the private sector, such as through Public Private Partnerships.

Simultaneously, there was the rise of another neoliberal concept, that of New Public Management, which was intended as a broad set of public sector reforms based around adopting corporate management ideals, at least in terms of using greater market responsiveness to guide public sector activities. Decentralised control, sub-contracting to private suppliers and setting performance targets and assessment exemplified these changes to public sector practices. As such, there was a new interest in the rules, institutions and practices of governing urban transport.

Why Governments Intervene in Urban Transport

Although the era of governmental reform to urban transport under neoliberalism sought essential changes in the place of government in the economy and, by implication, in community and individual life, there was often little appreciation in policy debates of the rationale for the original governmental involvement in urban transport. Indeed, even within many of the transport textbooks on the subject and those on urban policy reform, such a rationale is not provided. Some advocates of neoliberalism took it as an article of faith that governments would seek to expand their influence into urban transport; there is, however, a set of specific and essentially universal motivations behind the intervention of states into urban transport that have little to do with the inherent 'character' of governments.

These motivations can be described using a set of economic concepts, based on a central economic proposition that there are outcomes that markets can facilitate and outcomes that markets cannot, the latter being a condition known as 'market failure'. Ideologically, there are many reasonable objections to the starting proposition that markets should be seen as central to social life, but for the time being and for the sake of this argument, these will have to be put aside. Where market failures occur, achieving desirable outcomes and/or avoiding undesirable outcomes necessitate action by government or civil society. Even the most ardent market advocates accept market failure and the need for government involvement in economic and social activity; such advocates, however, set a higher bar for market failures and therefore seek minimal state intervention. A great deal of the normal operations and functions of urban transport in developed nations has little to do with free markets. Many developing nations with weak government exhibit the chronically failing urban transport created by free markets; such conditions never improve or evolve until governments (or well-organised civil society) intervene. Three types of market failure exist in urban transport and require state (or possibly civil society) intervention to resolve: collective goods problems, externalities and natural monopolies.

Collective Goods

Leaving corporations to develop urban transport according to their own wishes produces mobility services that will only suit a fraction of the urban population in the resulting pattern of over-serving (i.e. an excess of service providers for a given market), under-servicing (i.e. a deficit of service providers for a given market) and unserved locations. Over-servicing of popular routes results in excessive competition and fare cutting; this temporarily benefits consumers but results in corporate bankruptcies and industry instability, with the travelling public and failing corporations asking governments for assistance. Under-servicing and an absence of services results from neglecting areas and services offering comparatively lower returns to firms, so that many locations and times are without services.

Problems of this sort arise from the condition when resources are *collective goods*, meaning that it is impossible or impractical to restrict use of the resource; if it is available to one, then it is available to many. (Confusingly, collective goods are also known as 'public goods', but here we avoid this term because of its alternative meaning to denote things that are in the best interests of the general public.) Urban transport is not a perfect collective good because, past a certain point of use, crowding or loss of amenity occurs with additional consumption; in the real world, there are not that many goods and services with economic worth that have an unlimited capacity to meet demand for consumption.

Governments seeking to prevent or reverse such free market outcomes must intervene either by providing services through government agencies or setting rules to ensure corporate compliance with government intentions. Either dissatisfied travellers or firms can prompt government into action. Essentially, such government action seeks to ensure public benefits, usually that of the nominal goal of universal mobility (or something approximating this goal, subject to practical limits). Governments have a variety of motivations for such intentions, such as the ideal of providing equitable access to transport across social and economic differences. Voters in representative democracies may seek such services, requesting that their governments ensure that parts of the city are not deprived of public transport. Economic productivity and general public welfare can also benefit from universal access to public transport, as mobility is often associated with higher individual opportunities for production and consumption, such as increasing access to jobs and services; stakeholders in business and commerce can make similar arguments from their perspectives. Since under a free market for mass transport, the market will fail to provide universal coverage of public transport, governments act to ensure that the collective good of universal mobility through public transport occurs.

Externalities

A related but different type of problem that requires governments to act in public transport springs forth from a set of costs and benefits for which free markets do not have a response (i.e. *externalities*). As with the problem of collective goods, the problems of the costs public transport systems imposed on society and the environment became apparent quite early in the evolution of modern urban transport. Most obvious of these were the risks to public safety of trains and trams and the pollution from coal-powered vehicles in the crowded and growing cities of the nineteenth century. Governments also wanted to avoid the lost opportunities arising from fragmented and disjointed services using different equipment and with separate operations. There were financial incentives for corporations not to act to limit these costs and so governments intervened. Similarly, governments sought to capitalise on certain potential benefits that corporations would not provide if left to their own devices. Normal markets do not put a direct price on transport users for the benefits of road safety or costs of road trauma, the costs of pollution from vehicles, do not require vehicles to be roadworthy or drivers to be licenced, nor to have road laws or have incentives to seek economic benefits of an effective public transport system beyond those that directly benefit participating firms.

For communities and their elected governments to avoid these costs and to reap potential benefits that were not part of normal markets, governments had to change the rules and intervene in markets in various ways. State intervention on externality problems in transport takes many forms, including regulation of corporate activity to limit the causes of these costs (e.g. pollution control regulation, safety legislation, and regulation of vehicles and operators), taxation on undesirable outputs (so as to give the externality a market value) and public investment to capitalise on lost market activities (such as advisory bodies, research and setting standards).

Natural Monopolies

Another type of market failure in urban transport involves monopolies. This problem also emerged early in modern transport, such as when many horse-drawn carriages carrying passengers crowded city streets in the latter nineteenth century, creating competition to use the limited amount of public roadway. With the advent of horsecars and other vehicles using rails on city streets, the problem of monopoly transpired. Much public transport and transport infrastructure can only be feasibly furnished by a single supplier because of the singularity of the necessary infrastructure (meaning that it has no alternative uses), its expense, the extent to which sunk costs are difficult to recover and the impracticalities of sharing it between competing service providers (notably with fixed rail modes). For this reason, urban transport is said to have many *natural monopolies*. Economists have additional explanations for natural monopolies: they occur in markets with significant economies of scale, so that the costs (per unit) of providing additional goods and services decline significantly as scale increases. Although partly an issue of technology, economies of scale are often caused by the need to invest in large capital outlays – in this case, in transport infrastructure and rolling stock for public transport. These high capital costs pose a major barrier to the entry of rival suppliers, so that the first established in the market usually assumes a dominant position. Because natural monopolies have high fixed costs (and relatively low marginal costs), the monopoly supplier requires high levels of demand. Once a monopoly supplier is established, however, a potential market competitor is unable to achieve such low costs (as they are without the benefits of the economies of scale). Public transport conforms fairly broadly to these circumstances. Other providers of public mobility can potentially enter a monopolised market by having highly differentiated (i.e. specialised) services, seemingly countering the natural monopoly condition. Such rival services can only be competitive by offering specialised services within restricted markets, which usually leave the monopoly service largely unaffected (which is what we observe in practice).

Corporate monopolies can be assumed to abuse their power in economic markets and the history of modern urban transport is replete with examples. Abuse of market power includes setting unreasonably high fares, imposing prejudicial rules for using services, disregarding public safety and unreasonably low levels of service provision. Options for governments dealing with transport monopolies are relatively limited: regulating competition by law, using oversight institutions or by assuming ownership and creating a public monopoly. Governments acted to control market structure (i.e. public or private ownership), competition, prices and set rates of return. As governments continued to invest in urban transport, the strength of their monopoly position increased, but so too did the burden of ownership. Government investment strategies assumed a variety of forms, but transport services in the form of road infrastructure and public transport were rarely designed and organised in ways that provided income streams sufficient to cover their capital and operating costs. Public enterprises also appeared to suffer from dis-economies of scale when increasing size seemed linked to declining efficiency. Further, urban transport was usually the beneficiary of public sector cross-subsidy from other sources of public revenues (added to which there was usually cross-subsidy within transport operations, notably in public transport between different locations and services at different times). These problems of public sector debt and subsidy helped shape the debate over introducing market-based approaches to financing and owning urban transport (such as creating competition and breaking down the monopoly of public ownership, introducing privatisation and reducing government intervention and regulation of private firms).

Facilitating Progress

There is also a more intangible reason for governments intervening in urban transport and one that emerges strongly in the historical accounts, namely the ethos of progress. Governments intervened in urban transport because there were things they wanted that corporations could not provide, but behind this motivation was an articulation of the role of the state as organiser-in-chief of progress. Now there is much attention given to government failings in urban transport, but modern urban transport systems in the developed world are achievements of the state responding to the technological, social and economic opportunities of earlier times. Governments were uniquely qualified for this role, but by assuming control over urban transport, the states' role, reach and authority were also expanded. In effect, governments were responding to an intersecting set of issues that arose because transport systems constituted collective goods. Not only did governments have the singular capacity for financing the large-scale investments required, but they could also incur the financial risks of investing in such complex systems. Such a rationale was applied not only to public transport, but also to other utilities that also began as disaggregated, under-invested enterprises providing essential services to industrial societies. Overall, governments had a multiplicity of reasons for assuming control over transport; we need to recognise this overall rationale as one founded in a particular historical context and that this was subject to future change.

Governments' Role in Urban Transport

Government and urban transport engages all tiers of government and arrangements differ between nations considerably. Political institutions governing urban transport can be vested in the nation, state/province, territory, local government, metropolitan region, county, city, town and district. Most jurisdictions will have relationships with institutions at most of these scales. To these jurisdictions can be added the many varieties of regional or coordinating transport institutions, such as regional planning bodies and cooperative transport service arrangements. Across all the transport-related institutions of government is a great and often bewildering array of activities, ranging from the policy functions that regulate industry structures to regulatory regimes and specialist institutions and agencies delivering national advice and information, to providers of public transport services, including services such as selling tickets, cleaning trains and stations and providing security.

City and state governments found a variety of ways to intervene in urban transport to achieve their desired goals, with the resulting institutions and arrangements tending to be determined by local happenstance and national cultural approaches to government, as much as political ideology and political machinations within government. Generally, the responsibilities in urban transport cover transport planning, finance and investment, infrastructure provision and operations, road management and use (including driver and vehicle licencing, parking and active transport), public transport (including regulation, management and operations), freight transport, links with other public functions (such as land use and urban planning), public transport safety, social welfare and special needs transport provision and transport environmental impacts. Broadly speaking, these functions can be grouped into public transport and roads-related transport. Within state and city government transport authorities there are many functional arrangements, covering strategic planning, infrastructure planning and projects, service planning, road use and management and public transport functions.

City and state transport responsibilities are under the auspices of national transport responsibilities (excepting city states, such as Singapore), covering such matters as national transport policy, as well as technical regulations such as vehicle, fuel and emission standards; foreign ownership; import policy; federal support for transport investments; research and development; national taxation; national industry policy; national safety policy; standards for infrastructure design and construction; information gathering and data system management; and national transport projects.

Government involvement in urban transport engages many departments and agencies and many activities require coordination and liaison between spheres of government. Another complicating factor is that urban transport systems are networks with many sub-systems and integrations with other transport and urban infrastructure systems. Urban transport is usually part of a regional and inter-city transport system and the transport systems of large or strategically placed cities are connected to international transport systems. Managing these connections to other transport systems adds a layer of operational and institutional complexity.

Within this vast array of responsibilities, there are those that cannot be achieved by any other institution than government. In the roads-related realm, planning clearly exemplifies a state interest, whereas design, construction and management activities are more equivocal in terms of whether these can be privately or publicly undertaken. Similarly, within public transport infrastructure, the planning and design functions are government responsibilities, whereas construction need not be; for public transport services, planning and regulation are government responsibilities, but operations can be public or private.

Urban public transport provision by government spans the gamut of involvement. At one end of the spectrum is the 'full service' of state ownership and provision of all public transport operations. Such responsibilities can be any level of government: city, regional, state or provincial, and national. A lesser, but still substantial, level of government involvement is the state provision of facilities and services that might be used by state or other service providers. These services include roads, railway networks, stops and stations, information systems and ticketing systems. Towards the other end of the spectrum are governments with largely oversight responsibilities, with private operators providing transport services under a variety of agreements with governments. In this case, governments determine public transport service requirements (covering such matters as routes, fares and level of service) and monitor the performance of the private operators. There is also a minimal government role, in which most of the activity is undertaken by (largely autonomous) private providers, with governments approving fares, routes and levels of required service.

Transport Governance

Invariably, those cities suffering with the worst transport systems are those plagued by an assortment of failings in all forms of urban services, as they share a common root in having weak governments and poor governance of urban services. Market failures are also rife in these circumstances, as fair and effective markets cannot arise without being secured by a viable legal and regulatory system. As described above, the problem of collective goods and challenge of providing for the public benefit requires governments to intervene in urban transport. However, as the variety of urban experiences demonstrate, the actual role that government assumes in transport varies greatly as there has been considerable scope for private enterprise, and to some extent, community enterprise and initiative. Political and public policy inquiry has taken up these phenomena more broadly in recent years in an effort to reconsider the state's role and activities and whether and how some responsibilities of governments can be assumed by other agents, with the concept of *governance* entering into wider currency. Long-established assumptions of where the boundary between public and corporate should lay in the transport portfolio have been challenged; furthermore, unitary models of public sector governance have given way to more fractured, complicated and specialised roles for both public agencies and corporations.

Decision-making processes have also changed, with the traditional model of elected representatives in government making binding decisions implemented through public administrations by bureaucrats being replaced by a network of public and private policy actors involved in both formulating and implementing public policy from within, or as part of, legitimate state institutions. There is debate over whether governance concerns a new form of politics (because new 'non-government' actors can exercise political power), a new political institution (as a new set of rules and practices is in place), a new form of policy implementation (as governments choose between a range of different sorts of policy instruments) or refers to governments' influence achieved through a variety of non-hierarchical and non-coercive means (see Trieb *et al.* 2007). In practice, governance embraces the elements of politics, polity *and* policy practices.

This was also the era where managerialism had come to replace the older models of public administration. Public administration exemplified the idea of bureaucracy, where public officials play the central and hegemonic role in formulating and implementing public policy, following the rule of law and creating and enforcing rules and guidelines in a system that clearly distinguished between the spheres of government politics and administration. Managerialism came to public administration from the corporate sector with a basis in the older ideas of scientific management and featured giving greater and more discretionary managerial powers to senior managers, professional management expertise being valued over specialised experience, greater flexibility in operations and disaggregating units to form decentralised associations with corporate goals.

Facilitating the changes to governance by the state required of these new politics and circumstances necessitated reform to public administration, which took the form of New Public Management (Hood 1991). It was a marriage between rising managerialism and new institutional economics. For Western nations, the long-standing models of public administration reached their apotheosis in the rise of the welfare state in which public administration assumed responsibility for a comprehensive delivery of social and economic policy. Towards the end of the 1970s, traditional public administration was undermined and reformed by the New Public Management movement, changes that were politically associated with the rise of neoliberalism (i.e. neo-classical economics and associated rational/public choice theory), notably in the English-speaking OECD nations. Rhetorically, the themes of this revised view of public administration featured terms such as leanness, competitiveness, value-for-money, transparency, accountability, flexibility and responsiveness. Hood (1991) linked New Public Management to four trends of the era: 1) Efforts to curb or reverse the growth of government spending and employment, 2) Privatisation, 3) Automation, and 4) Development of an international agenda focusing on general issues of public management.

In short, these changes saw management practices of business applied to public policy implementation and service delivery under the rationale that efficiency in resource use and improved efficacy in service provision could be best achieved using the insights, incentives and practices of corporations in market settings. The traditional public service ethos was replaced by beliefs in the benefits of markets and role of competition. From an organisational perspective, public administration went from being essentially a closed system to one that was open and responsive to market influences. Changes were wide-ranging for both service delivery and for public administration functions. Within the public sector, there were new managerialist approaches with increased managerial autonomy, performance monitoring and reward incentives, together with new systems of accounting, auditing and finance modelled on private enterprise. These functional changes were made with a view to increasing attention given to the internal performance of the organisation, achieving organisational goals and on achieving desired results. Internal reforms to government agencies under the dictates of New Public Management usually occurred in concert with neoliberal reforms to the government's role in the transport portfolio, thereby forming new systems of governance. Neoliberal changes included greater use of contracted services, fixed-term employment, replacement of permanent employees with causal labour, allocation of resources using market principles, cost-cutting and service rationalisations, but also greater responsiveness to consumer demands, increased marketing and use of markets in a variety of ways (as discussed in this chapter). As a result, existing governance institutions were altered or eliminated and new institutions created. Significantly, some state functions were given to new entities that operated at arm's length from government, particularly those with economic development roles and associated with major projects, sometimes reporting directly to government and not through the established government departments.

We can find a wide range of examples of these changes and the rise of governance as the state goes from being a unitary provider concerned with policy and its implementation to having a regulatory role and a focus on managing resources and performance. Trieb *et al.* (2007) point to the emergence of soft laws and guidelines in public policies as examples of governance, where governments decide not to use binding legal instruments in deference to stakeholder interests. Similarly, the authors refer to the growing use of flexible approaches by governments, giving policy actors latitude in deciding how to conform to policy requirements or offering choices in the ways in which conformity can be achieved. Governance also implies that some governing processes are carried out by private groups, such as self-regulation by corporations, groups comprising government and corporate membership or having policy implemented by corporations. As a result, such governance often occurs through networks of policy actors, rather than through the fixed hierarchy of government, so that decision-making powers and political authority becomes more dispersed compared to the centralised power of governments. In some respects, movements such as 'smart regulation' are part of these changes, with the shift from government regulation to self-regulation by business and third-party oversight, such as by public interest groups.

Governance in developing nations is often considerably weaker than in developed nations and developing nations have a wide range of governance arrangements, reflecting local cultures, colonial influences, international development influences and local modernisation efforts. Developing nations often have very large cities and national governments may play a role in urban transport decisions that might usually be left to city authorities, creating political tensions and deficiencies in transport planning and ultimately in service delivery. Existing problems of low household incomes and few public resources for transport investments tend to be associated with poorly developed transport and other infrastructure and a lack of expertise, particularly at the local level. By default, cities with weaker economies with weaker planning tend to have a greater market orientation and reliance on the private sector.

There are a number of implications around these changes for the goals of sustainable transport and these are considered later. However, it may be useful to briefly consider the role of governance in urban transport. Essentially, there are four forms of urban transport governance: private transport, public transport, informal transport, and as argued in this volume, community-owned transport; the first three are considered here and community-owned transport is considered in its own right in a separate chapter. These aforementioned developments and arguments over governance have influenced the role of government, the shaping of institutions to exert governments' will and how the roles of institutions are understood and applied in practice to urban transport.

Urban Transport Institutions

Understanding institutions broadly means considering them as social systems of rules and structures that are linked or tied to particular behaviours and procedures for social purposes. Although transport systems were, and still often are, considered as infrastructure, as Stough and Reitveld (1997, p. 207) point out: 'There are however, a number of institutions ranging from laws and regulations to informal conventions that support transportation systems. These institutions are defined by and maintained at base by culture and values.' Although this cultural foundation provides institutions with considerable stability over time, rapid external changes, often through technological change, produces 'rapid and significant change in culture, values and institutions' (Stough and Reitveld 1997). Today's quest for sustainable transport, with its new environmental, technological, equity and economic challenges is the most recent of these transformations in urban transport. Sustainable transport is being sought in the context of an array of several other changes underway. Understanding the changes underway in contemporary urban transport requires knowing something of what is happening to transport institutions and also of institutional inertia that provides resistance to change.

A key economic and associated institutional change underway is the shift from Fordist models of industrial production to the contemporary post-Fordist condition. Fordism featured concentrated production in fixed locations, large scale (because of the benefits of lower costs available in mass production of standardised products) and vertical integration of the phases of fabrication, manufacture and distribution. Once the ideal of all major producers, Fordism has given way to more flexible production, with processes and activities disaggregated, decentralised and outsourced and greater attention given to customised outputs for markets with quickly changing preferences (Harvey 1989). Management structures have followed suit, moving from centralised and hierarchal organisations to more dynamic, flexible and adaptive approaches.

Governance has had to evolve in response to these changes. Long-standing and stable institutions could function effectively in environments that were predictable and steady, but existing regulations and procedures became ineffective and burdensome in the more dynamic and unpredictable post-Fordist world. Responding by stripping back governance to simpler structures was not an option; rather, the increasing scope of competing interests and expectations of urban transport under such paradigms as sustainable transport added to the expectations of government. Expansions of these domains brought new stakeholders into the political processes for decision-making and challenged existing institutions. Folded into this mix has been the rise of neoliberalism in transport governance, so that the traditional roles of states in transport are being reformed in line with advocacy for greater use of market mechanisms, reduced government activity and expanded spheres of corporate involvement.

Public sector agencies in urban transport have been similarly transformed along with the wider economy in developed nations. In many instances, the agencies managing public transport had a clear lineage to the origins of state involvement in urban railways and streetcars in the 1800s and for road transport agencies, to the first quarter of the nineteenth century. Stough and Reitveld (1997) state that the structure of these agencies was Fordist, being 'somewhat inflexible, vertically organised and insular', following civil service codes and with professional workforces dominated by trained engineers. As privatisation, outsourcing and other changes have occurred, these public agencies have also been transformed by being more flexible, integrated with other agencies, more autonomous and using the new management technologies. Often, these agencies have also increased communications and media outputs to market agency activities to a more diverse community of users.

Significantly, the workforces of public sector transport agencies have changed from being highly unionised, regimented and securely employed to something closer to the neoliberal ideal. Short-term contracts, no union membership or under more tightly proscribed unions, employment by corporations rather than public agencies – the transport service workforce in many cities differs from the model that existed throughout the last century. Following the post-Fordist model, many of the public transport functions that were with a single agency have been split into functional units and many functions outsourced, such as accounting, advertising, customer relations, information technology services, marketing, media relations, station cleaning, station staffing, training, vehicle maintenance and vehicle cleaning. Job loss was often part of these changes. Union unrest and industrial disputes have frequently been reduced as part of these changes and this prospect was no doubt one of the motivations for decision-makers.

A complicating trend is that of calls for greater integration in urban transport. Efforts to have increased modal and service integration in public transport within cities has often required institutional reform to ensure such goals as connecting services across urban systems, coordinated timetables, common ticketing systems, and single advice and customer communication systems are fulfilled. Economic integration between cities also challenges traditional institutional arrangements, such as results from new international trade agreements and supranational entities, of which the EU is an obvious example. In these circumstances, national policies and national agency responsibilities may be revised in light of a new international authority (covering, for example, trade, international standards and taxation). Even more mundane connections, such as the need to coordinate services between cities with city services can require new arrangements and agreements. While the traditional models of public administration are now largely superseded by new institutional models and arrangements, this evolution will continue to be shaped by external circumstances and changing political and social demands.

Public and Private Transport Systems

Considering the realm of public and private transport rests on a considerable assumption, namely that there is an effective system of government in place, for without the active role of the state, transport is unregulated and gives rise to essentially informal transport and a range of market failures. Across the world there are a variety of different organisational arrangements for urban transport; all retain some role for dealing with the collective goods aspects of transport, but what varies is the extent to which corporations and markets are used. While there is a public policy debate over the extent which the role and functions of government can be replaced by corporations and civil society, there remains a set of core state functions as outlined above that cannot be reduced without compromising the functions of modern transport systems.

There are various ways of determining the identity of the public and private sectors. Karlaftis (2008), for example, used these attributes: ownership of assets, regulation by the public sector, system management, extent of private sector involvement and public subsidy. Arranged in a continuum with public and private ownership at either pole, the differences in identity appear clear and familiar at the extremities. Public monopolies feature full public ownership and a closed market that is fully regulated, with public agencies responsible for operating and managing the transport system, and public subsidy of operations. Liberalised markets for transport are at the private ownership end of the continuum, characterised by having several owners who compete in a deregulated market for service provision and although there is private ownership of management and operations, there may be public subsidies. Limited competition or franchise models sit somewhere between the fully public and the liberalised markets and can have public or private ownership, use competition to determine who can participate in markets, have somewhat relaxed regulation, with public or private operations and management and usually receive a public subsidy.

A curious feature of public transport is that, despite its high profile and familiarity, its identity is both assumed and uncertain; many textbooks on urban transport and public transport are without a definition of public transport. Approaches to understanding public transport include definitions based on mode (i.e. vehicles carrying multiple passengers), access (i.e. transport services available for all users), ownership/property rights (i.e. state-owned services), service type (i.e. mobility services with schedules and fixed stops), passenger volume (i.e. mass transport provision), market-based (i.e. services provided in closed or regulated markets), political (i.e. electorates hold governments responsible for public transport services) and law (i.e. legislation determines roles and responsibilities of governments and corporations in service provision). Many definitions bundle several of these themes, such as Walker (2012, p. 13): 'Public transit consists of regularly scheduled vehicle trips, open to all paying passengers, with the capacity to carry multiple passengers whose trips might have different origins, destinations, and purposes.'

Public transport is therefore defined according to circumstance and the stakeholder's interests. It is not that most of the approaches used are necessarily incorrect (although some have undoubtedly been poorly thought through), but rather that the meaning depends on the intended application. In practice and in theory, the identity of public transport can be difficult to establish: most urban public transport contains a great number of corporate relationships and indeed, even state monopoly services may be provided by state-owned enterprises. However, the problem remains that some difficult public policy questions, such as the limits of privatisation, appropriate use of market-based policies or the identity of taxis and informal transport, require greater inquiry into the character and identity of public transport.

Historically, before the advent of mass transport and accompanying government intervention, there is effectively no public urban transport as understood in the modern world. As urban and regional transport develops, and when states began to assume a major role in organising modern transport systems involving motorised transport in the late nineteenth century, the idea of public transport became strongly associated with transport services provided or controlled by governments at the local, regional, state, inter-state and national scales. Of course, the role of central authorities and states in providing for roads and other transport infrastructure dates back to antiquity, while the idea of ownership by the state is associated with the rise of the modern state (i.e. post-Westphalian states). Such state intervention has come to assume many forms, ranging from government entities, public corporations, government coordination bodies, mixtures of public and private enterprises, public management of state-let contracts and franchises, to varying extents of regulation of private operators.

For our purposes in this volume, governance and associated public institutions can be used to understand the identity of public transport and, importantly, cast a clear divide between public and private transport. Three governance criteria define public transport: 1) Governance of the service or related activity is through public policy mechanisms, 2) Financial structures for the transport and related services are based in public agencies, and 3) A primary objective of the system operators is providing a transport service. Public transport is the responsibility of governments who exercise their will through public policy, as well as the service providers that can either be government agencies or firms with government contracts or franchises. Private transport is also subject to laws and regulations, but crucially, while the chain of accountability for public transport ends with the government, for private transport accountability ends with those owning the corporation (which could be private or through shareholders and boards). Public transport, from the perspective of the state, has as its goal providing mobility services, as opposed to firms providing passenger transport that are determined by corporate goals, such as profit, shareholder demands and market share. Governments are held by social contracts with electorates over public transport services, so while state-owned enterprises might also pursue corporate goals, governments remain the ultimate owner of public transport and burdened with inalienable public service obligations. Essentially, it is possible to consistently distinguish between the public and private realms in transport systems based on the system's features of governance. While there might be many definitions of public transport, private transport is taken as being entirely or largely self-explanatory. Again, adopting the perspective of governance, private transport covers infrastructure, transport services and vehicles under private ownership for private use with the locus of governance within the private sector.

Informal Transport

A distinctive sight in the streets of many developing world cities is a fleet of small, independent, passenger-carrying vehicles, collecting passengers from the streets, operating without set timetables, routes or fares, especially in low-income areas; this informal transport sector (aka 'paratransit' and 'Intermediate Public Transport') provides an essential mobility service where public transport is absent or severely wanting. A wide variety of names are given to these services, including angkots, bemos, jeepneys, matatus, minibuses and tuk tuks. As Cervero explains (2000, p. 5): 'The informal transport sector comprises mostly small-vehicle, low-performance services that are privately operated and that charge commercial rates to, for the most part, low-income, car-less individuals making non-work trips.' 'Informal' does not cover all unregulated services, such as offering people a lift for free, as informal transport is a commercial service.

There are many views as to what constitutes informal transport; a crucial and distinguishing feature appears to be that it refers to mobility services that are without the official (i.e. legal) sanctions that public and private transport enjoy. Informal transport is also generally without any supporting infrastructure. As described above, urban transport is highly regulated as a market and as a system it is tightly controlled and managed by state authorities. For passenger transport providers, there are rules and regulations covering vehicles, their design and use, who and what they can carry and not carry, where they can go and not go, taxes applied to roadways and vehicles and licences for carrying passengers. Operating largely outside this regulatory web is the realm of informal transport: these operators are usually without permits to operate, their vehicles are without permits to carry passengers, the drivers are without official licences to drive passenger vehicles, there are no service standards, fees are negotiated rather than being set, there are no formal service territories and the vehicles and businesses are probably uninsured (Cervero 2000, Cervero and Golub 2007).

Informal transport performs a critical function in developing nations, often providing mass mobility and, as such, is usually tolerated by public officials. In many cities, the bulk of motorised and much non-motorised mobility is provided by informal transport. These services provide on-demand mobility, meet niche mobility market demands, fill gaps in transport networks, can be a significant source of employment and obviously generally provide services where these are largely deficient (Cervero and Golub 2007). Public transport in these cities is often hampered by chronic under-investment, with old and unreliable rolling stock and buses, services can be inadequate, road-based services are stuck in traffic and overall revenue streams tend to be comparatively small (Dimitirou and Gakenheimer 2011). Further, the overall transport infrastructure is often not sufficient to meet total demands, giving rise to congestion, sometimes on a massive scale. In developed nations, informal transport is a fringe activity that carries few passengers overall and receives no special treatment from ruling authorities; if caught, informal transport providers are at risk of being put out of business and the owners/operators, prosecuted by the courts.

Informal transport represents mobility in a near free market setting, i.e. laissez-faire transport. In this setting, we gain a perspective of the interplay of market forces on mobility services and also appreciate the necessity of governments for regulating the behaviour of transport providers. Cities relying on informal transport are typically plagued with mobility problems (Luthra 2006, Kumar *et al.* 2008). Informal transport operators are infamous for a range of sharp practices, such as blocking rival operator's services, intimidating rival operators, turning around to collect additional passengers, racing rivals to reach passengers and waiting for vehicles to fill before leaving. Informal transport can be a major contributor to road congestion and contributor to air pollution. Conditions for travellers and other road users are often openly dangerous, with drivers caring little for passenger safety and comfort, and driving aggressively in crowded thoroughfares and violating road laws.

Developing world cities without coherent transport policies and effective institutions often resort to a passive regulatory approach with rigid bureaucracies and uneven performance, combined with certain *ad hoc* arrangements where public officials assume responsibility in conferring various concessions and allowances (such as the right to provide transport services and infrastructure). Informal transport operators have few resources, offer minimal services and seek to secure local markets though legal and illegal tactics (Meakin 2004). Such a combination encourages informal transport as a solution of last resort, but one that is invariably prone to corruption of officials and other illegal practices by operators. One strategy by the operators has been to form collectives that generate a greater capacity to negotiate with officials, but can result in market abuses in price-setting.

These factors interact to produce a system that may appear chaotic, but within this chaos is a stability that prevents the development of superior arrangements for improved passenger mobility. A persistent concern has been that informal transport undermines public transport in developing nations with cheaper fares and direct services, making the task of public transport more difficult. Informal transport is not public transport and exists largely outside the formal transport system. Governments have been pursuing two strategies to deal with informal transport: 1) Finding ways to bring informal transport into a regulatory system, and 2) Integrating informal transport into existing public transport systems (World Bank 2005; Hook and Fabian 2009). One issue discussed later in this volume is the rise of new carsharing and ridesharing businesses around the world; whether or not these are a type of informal transport is canvassed in Chapter 4.

Urban Transport as Infrastructure

Transport history tends to concentrate its attention on the technological developments in vehicles, so that the usual narrative concerns a succession of technologies of allegedly superior character. A key feature of urban transport is that vehicle technologies are but a relatively small part of a complex technological system supporting the major types of vehicle, including such elements as special infrastructure, energy sources, energy distribution systems, vehicle manufacture and maintenance, regulatory and legal support and control, research and development, control and management systems, as well as expert knowledge and training. Developments in transport technologies are necessarily complicated transitions involving urban infrastructure systems. When horse-drawn trams were put onto rails, for example, the necessary tracks were laid down in the streets of popular routes and the trams, other vehicles and pedestrians competed for road space. With motorisation, there is a need for other technological developments, including specialised infrastructure. Critically, without these infrastructures, the emerging motorised vehicles could not have succeeded in the ways that occurred.

Infrastructure and associated technological developments did not occur deterministically because of the popular endorsement of new forms of mobility, but also because a number of other associated developments occurred at the right times in history. Looking at the different times technologies have been taken up in different places around the world provides some validation of this point. One feature of the rise of dedicated transport infrastructure is the growth of government in the western world; more complicated and extensive transport systems can only be realised through expansions in government authority and administrative competency. Large-scale transport investments in road and rail are expensive and governments were willing and able to undertake such investments. This confidence spoke to the strength of government institutions, such as the rule of law and the role of state as an unimpeachable source of credit. Expansion of government powers and public expenditure was linked to the growth of taxation, including corporate taxation. Expectations that democratic governments of elected representatives would, and should, be 'nation builders' are directly tied to the growth in government income and resultant expenditures on investments of state and national importance. Following the Great Depression, the US's New Deal and the widespread adoption of Keynesian policies for state investment in national economies, governments of Western nations played a greater role in promoting economic activity through the twentieth century up to the monetarist era of the 1980s. Stronger governments were also important in providing conditions for political and economic stability, without which the modernising ideals and the willingness of capitalist institutions to invest on a large scale are soon lost.

At a more practical level, the early 1900s saw a new era of urban and transport planning emerge in response to the needs of cities growing under the influence of new forms of mobility. Prior to the industrial city and even during the Industrial Revolution, there were no surprises that ideal urban planning would draw strongly on the classical cultures of Greece and Rome (Hall 1988). These new forms of mobility were creating new demands on cities for the planning of transport and development that exceeded such classical outlooks, ushering in new visions for urban planning. Governments had an increased capacity and interest in considering transport as a means to deliver social and economic goals; this was entwined with a new capacity for setting strategic goals for developing the state and nation. On the ground, such planning began to take into account the design and function of transport systems, involving such factors as demand forecasting and capacity estimation, so that the design of the system would meet expectations.

This growth of state power is not an argument that land use and transport planning were entirely successful or that state authority was rationally directed to ensure optimal development of the transport system. Indeed, the critique of environmentally sustainable transport questions the entire project of automobility that dominated state attention of most developed nations through most of the last 100 years. Examples of failures in transport planning abound; for instance, Peter Hall's (1980) *Great Planning Disasters* includes a chapter on London's motorways. However, the basic point stands, which is that the design, funding, approval, construction, management and use of major urban transport infrastructure is a function of state authority, legitimacy, and political and economic power.

Urban transport somewhat fits into the usual demarcations of technological development, but not quite. Through the Industrial Revolution and the age of the railways, cities are connected to each other, but despite many experiments and attempts, it proves difficult to apply railway technology for mass transport within cities. At the height of the Industrial Revolution when industry is being transformed, the horse and walking are the primary sources of urban mobility. Urban rail and subways certainly exist during the age of the steam locomotive, but it was the successor railway technologies (mainly electricity) that made railway use within cities viable, i.e. in the Second Industrial Revolution. In this phase, after about 1890, electricity is also used to move trams, finally replacing the horse and mass transport becomes far more widespread within cities and across the industrialised world. Private motor vehicles, an invention from the Second Industrial Revolution, do not become significant until after this period concluded.

As Graham and Marvin (2001) describe, cities are being transformed from assumptions about uniformity to an increasing fracturing and diffusion as a function of massive networked and inter-related and connected infrastructures. After WWII, they argue, an ideology was established for infrastructures that were considered to deliver somewhat universal services for all, or nearly so, at uniform costs across different cities and areas. Infrastructure networks were 'widely assumed to be integrators of urban spaces', so that they integrate cities and regions into 'functional geographical and political wholes'.

Urban infrastructure, including transport infrastructure, however, is fragmented and divisive through privatisation, specialisation and customisation. Transport services are not uniform across cities and poorer areas often have weaker services than their wealthier counterparts as a consequence of economic markets, decisions by public officials or a combination of the two. These circumstances often prompt communities and entrepreneurs to meet local mobility demands through informal transport services. They also point to customising infrastructures for investment enclaves, such as those provided through public expenditure for private and exclusive use. Gated communities and closed private streets maintain self-contained communities through which there is no open or pass-through access; the management of such places is private. Pedestrian footpaths have also become privatised by providing access between commercial and other buildings separate from the public footpaths. Such developments are part of the fragmenting urban landscapes giving rise to clusters of 'packaged landscapes made up of customised and carefully-protected corporate, consumption, research, transit, exchange, domestic and even health care spaces' (Graham and Marvin 2001, p. 5). Transport infrastructure, rather than being a force of integration, plays a key role in this fragmentation and is itself a part of this fragmentation. Privately owned tollways exemplify fragmented transport infrastructure, operated for profit and designed to suit the needs of 'affluent commuters on particular urban corridors'. Private tollways provide alternative and quicker journeys than public roads.

As Graham and Marvin (2001) describe, between 1850–1960 in the Western world, there was a growth of networked and monopoly urban infrastructures, which came about with a shift from 'piecemeal and fragmented' systems to ones both more centralised and standardised. These authors note that adapted versions of these systems were spread into the colonial cities (and generally for elites) in Asia, Africa, South America and Australasia (Graham and Marvin 2001). As old cities were being remade through energy, water and wastewater, communication and transport systems, new suburbs were being constructed, as industrial cities were created out of older pre-industrial forms and new links between cities forged. Social life was also transformed as the forms and processes of life in industrial cities became universal, with the new infrastructures and associated technologies helping to orient life around industrial production, distribution and consumption. Underlying the creation of the infrastructure systems was the belief that a city could be organised and operated as a single industrial artefact. According to this view, by using science and technology, the various service requirements of the city as a whole could be rationally organised. By the time of WWII, these 'single, integrated and standardised' networks were completed in nearly all cities of developed nations, providing services to an extent and level that was unimagined half a century earlier. As Graham and Marvin (2001, p. 41) conclude: 'These were legitimised through notions of ubiquity of access, modernisation and societal progress, all within the rubric of widening state power.'

Furthermore, it may be better to describe the sequence of changes as development, rather than assuming that these changes are axiomatically a form of progress. In particular, we are interested here in trying to understand what problems these changes in urban transport were attempting to address and, conversely, which problems were of less interest. Graham and Marvin (2001) make two pertinent points: firstly, the difference between rhetoric and practice, the ideals of integrations and cohesion and their opposites (2001, p. 42):

> Beneath the universalising rhetoric, modernising cities were always about rupture, contradiction and inequality. Extending metro, sewer, water, highway, energy and communication grids through the fabric of city spaces was always laden with social and political biases, highly uneven power struggles and cultural and historical specificities.'

Secondly, that these effects were socially transformative: 'Notions of space and time, speed and culture, subjects and objects, technology and society, were gradually recomposed' (Graham and Martin 2001, p. 42).

Neoliberalism and Urban Transport

For one hundred years or more, the norm in industrialised nations was state ownership of public transport systems and for creating state bureaucracies to manage these systems. This eventually gave way to a sea change in the form of a turn towards market-based approaches for providing public transport (i.e. neoliberalism, as raised in the Introduction). Whether these changes have now reached their zenith is difficult to discern; some argue that this point has been passed (Docherty *et al.* 2004), whilst others see the process as continuing and evolving (Macário *et al.* 2007). A complete account of the world's public transport management systems is not readily available, although a number of reckonings underscore the wide variety of management systems in place, with great variety in the relationships between governments and firms (e.g. OECD/ITF 2008). Further state-owned systems are being subject to neoliberal reforms at the present time, but against this trend is also a smaller counter-movement of states reversing earlier market-based reforms and resuming control over public transport systems (such as transport re-nationalisation in Argentina). As a result, the overall picture is one of great complexity and any claims that neoliberal reform will eventually embrace all public transport systems must be regarded with caution. What may also occur is the emergence of new institutional forms that defy ready identification as being either neoliberal or state-owned.

Such is the scale and intricacy of urban public transport systems that there are a multitude of ways in which private firms can be engaged. Even before the wider publicity given to the neoliberal reforms in public transport, there had been a considerable degree of private sector activity in the form of contracted service operations. A number of urban systems regarded as providing best practice in terms of services offered have been using contracted private providers for operational services to state or city authorities (such as in a number of European nations). Debate over the virtues of the market-based reforms has been lively, with many contested claims over such measures as the cost of services, reliability and punctuality, influence on passenger numbers, levels of services, operational safety, staffing levels, innovation and investment levels. Objective generalisations are difficult to formulate on the basis of the experience of single jurisdictions because of the singular character of each system that reflects local conditions. Comparative analyses between systems can overcome the limitations of individual system studies, but necessarily reflect the selection of the case studies chosen and are often limited by differences in data collection in different jurisdictions.

Neoliberalism seeks to replace public sector services with corporate providers under the rationale that private sector performance is superior where an effective market exists. (Subsidiary privatisation objectives, from an economic perspective, include greater competition in the market, benefits to users in lower prices and higher levels of service, lowering of government debt and growth in infrastructure quality and capacity.) This credo is critical, for much of the controversy over neoliberal reforms has its roots in whether or not the services in question can be provided exclusively by government (i.e. those opposing neoliberalism identify the necessity of government because they see public services as necessary in conditions of market failure). Central neoliberal claims are that private enterprises provide services at lower costs and with greater efficiency than the public sector, passing savings onto the state (thereby reducing the need for public subsidy of public transport) and to consumers. Other claimed benefits are that reducing regulation also improves private enterprise performance without compromising service and legal obligations (such as safety, consumer welfare and environmental performance). Investing in infrastructure and rolling stock is more likely with private sector involvement, claim its advocates, as the public sector has often under-invested in the sector to its detriment. An example of superior private sector performance is the claim of greater responsiveness of private firms to consumer demands and preferences. Another key claim of neoliberal advocates is that reforms to work practices and wages of unionised workplaces can be better achieved by firms than by governments. Government decision-making in these and other areas is inferior to those made by firms, the neoliberal argument goes, because governments are influenced by a wider agenda of political factors that promotes economically inferior, if not irrational, outcomes from an economic perspective.

Neoliberal reforms to public transport have been, for the large part, a functional disaggregation of state management. There has been no change to the basic problems that necessitated state ownership dealing with natural monopolies, the need to capture the benefits from planning and coordination across the system and the requirement to deal with the costs that would arise from free market service provision. Where there are functions and issues derived from these market failures, we find that states have usually retained the full measure of the authority and control over the system. Accordingly, state authorities in systems that have undergone significant market-based reform typically retain control over planning for future growth, setting service requirements (such as over routes, stops and stations, frequency of service and punctuality), providing safety standards and have some kind of influence over access and equity issues. Problems of monopoly supply have been addressed (although not necessarily resolved) through such means as requiring competitive contracting bids, fixed-term contracts, performance-based contracts, performance monitoring, performance enforcement and sanctions. States set rules to govern interaction between service providers and often establish (or require to be established by the operators) specialist institutions for such tasks as integrated ticketing between providers/modes/services, marketing and publicity, service coordination and customer relations. Within this reform there has considerable internal evolution, with learning from early failures and missteps, and the sharing of experiences (e.g. Macário *et al.* 2007). New institutions are created by these changes, so that neoliberal reform has not been a return to the free market conditions of the nineteenth century, but rather the (re-)entry of private firms into a framework in which the state retains a strong and central oversight role.

Amongst the most popular approaches to the use of market-based instruments by governments are (Estache and de Rus 2000, OECD 2003):

- Performance programmes and contracts: These agreements between state and 'autonomous public enterprises' establish the services to be delivered by specifying outputs and performance levels over the period of the agreement (usually only a few years, but also usually renewable). Payments under these contracts are usually made to cover the costs of investments, rather than for operational costs.
- Management contracts: A fixed fee (plus possible incentive payments) is paid to the contract holder for management services of assets that remain owned by the government. Contract duration is usually of several years and governments retain the commercial risks of transport operations.
- Franchises, licences and concessions: There is no asset transfer under these agreements as the government retains ownership, but receives from the franchisee an agreed set of services from using those assets (such as operating the bus or rail service, but not owning the infrastructure or rolling stock). Sometimes the franchisee supplies additional assets of their own. Of longer-term duration, sometimes several decades, these agreements transfer service risks (or a portion of them) from the state to the corporation. There are usually subsidies to franchisees from governments.
- Service contracts: These agreements are small in scale, of short duration and are usually allocated through a tendering process. Rather than collecting revenue from users, the private operator provides a service for which the government pays. An alternative model is a net costs contract where the operator also collects the revenue directly.

Neoliberal political and economic theories have been widely adopted by OECD governments and by some nations in their transition from centrally planned economies to more capitalist economic forms over the past three decades or so. (Some of the ideological and philosophical aspects of neoliberalism are examined in Chapter 10 in the context of the political economy implications of community-owned transport.) A range of motivations were involved in the shift to neoliberalism, including changes in ideologies, concern over high levels of public sector spending and debt, technological changes that made monopoly ownership obsolete, globalisation of global financial markets and economic change following major political change (OECD 2003). Experts differ in their interpretations of the exact meaning of neoliberalism, but there is much basic agreement about its key features, remembering that this neoliberal activity has both political and economic dimensions. Amongst the OECD members, it is the governments of the English-speaking nations that have responded most strongly to these changes in adopting neoliberal practices, although arguably all have taken up neoliberalism to some extent (Harvey 2005). Under the influence of international agencies, notably the World Bank and International Monetary Fund, many nations in Africa and South America were encouraged to privatise their railways in the 1990s. This change marked a transition from common models of government built around strongly centralised state authority, extensive public provision of services and market regulation. Prominent features of these changes are such initiatives as the corporatisation of public sector activities, privatising state-owned enterprises, creating new property rights and opening up monopolies to competition.

Public transport and the public institutions involved in its operation, planning, management, regulation, ownership and state oversight have undergone considerable changes in industrialised, democratic nations. Public transport is therefore being shaped in response to these two major policy debates, that of its major role in promoting sustainable transport (as driven largely by progressive political interests) and that neoliberal institutional reform (driven largely by conservative political interests). It is notable that these debates have been essentially separate and there has not been much scholarly or institutional attention given to the relationship between these issues, such as to inquire into the sustainability implications of public transport privatisation.

Barriers to Change: Institutional Path Dependency in Urban Transport

Rationality has had a large place in understanding and evaluating institutions. Under rational models and normative expectations, institutions should trend towards greater efficiency over time. Deciding what efficiency exactly means can be an elusive quest and most explanations evoke physical system models, so that the ratio between a given input of resources and a corresponding set of outcomes defines efficiency. Public administration textbooks, for example, frequently depict causal chains of rational decision-making, with developing options, developing assessment criteria, option assessing, option implementing, performance monitoring (and subsequent feedback to further option development) as the rational process of institutional evolution. In practice, large-scale institutional change rarely follows the rational model (as established by social scientists, such as Herbert Simon and Charles Lindblom in the 1950s), but the underlying ideal of purposeful reform and development towards betterment is maintained as a primary directive for institutional change in response to a dynamic world.

But this does not always happen. A major problem in institutional studies (and in other fields of inquiry) is in understanding of why institutions become locked-in to certain practices, procedures, arrangements and technologies after more efficacious options have emerged. To employ a biological analogy, if the external environment alters sufficiently to favour evolutionary change, why do institutions fail to evolve and remain with the status quo to their own disadvantage? Clearly, the inertia of history is highly influential. Partly, of course, there is an issue with the formulation of the problem and the limitations of rational models in situations where political processes are in play; however, the issue of institutional inflexibility remains important in understanding why institutions can and cannot be transformed.

One description of the problem that has become popular is that of path dependency, especially in economics and political science. This concept is closely associated with a renewed interest in institutions and the new institutional economics, and the view that institutions influence behaviour and therefore influence outcomes, prompting a large body of investigative work into institutional design and effectiveness (especially as part of international development projects). Path dependency is not really a theory, as such, as it does not have predictive powers or provide testable hypotheses; it is more of a category or condition. There are several defining aspects. Systems with path dependency exhibit conditions that are self-reinforcing and build institutional stability. Over time, the benefits of existing arrangements continue to accrue and become more popular, so that the investment of resources, time, expertise and forms of social capital increases. This stability is also bolstered because there are perceived high costs of change; the more the existing conditions are strengthened, the greater the costs of change. Yet no conditions last forever and when change occurs or could occur in path dependent conditions, this happens during so-called 'critical junctures'.

Economist Brian Arthur (1994) considers these features to create conditions of 'increasing returns' in path dependency:

- Economies of scale: Large establishment costs inhibit any impulses to change, especially considering that unit production costs decrease with higher outputs.
- Improvements in productivity resulting from learning.
- Adaptive expectations: Increased uptake leads to less uncertainty about technologies (i.e. technological performance and reliability).
- Network effects that result from greater and wider involvement: 1) Benefits accruing from other's use of the same technology as promoted by multiple and interdependent technologies, associated infrastructures and interlinked institutions, infrastructures, economic relationships and technology, and 2) Larger networks become more attractive to more other potential users (i.e. there are benefits from technological compatibility).

Another point can be added to Arthur's list. Some systems persist simply because of the longevity of equipment and infrastructure; in some instances, the owner's original capital debts have been paid and they have only to meet the comparatively cheaper ongoing operating costs. When these operating costs are low compared to those of replacing obsolete equipment, then there is little economic incentive to replace such durable capital equipment. Uncertainty too is an overarching disincentive to change, so that the less the future can be predicted, the weaker the incentive to invest in new capital, procedures or products.

Path dependency has been applied to so many fields and problem types that it risks becoming excessively generalised. Confining our interests to two fields particularly relevant to transport institutions, the work on path dependency by Douglas North (1990) and Paul Pierson (2004) on institutions and Brian Arthur (1994) and Paul David (1985) on technologies are insightful (the latter of which has attracted greater controversy within economics).

Arthur and the school of new institutional economics view institutions through an economic lens; they argue that institutions exist because economic markets cannot be perfected. For economists, because of our imperfect knowledge, institutions (and property rights) are essential to create efficient transactions in the marketplace. Politics enters their argument, as the institutions in question are creations of ideologies and political processes, so that the performance of institutions determines the performance of economies. Further, as institutions are formed by those stakeholders with the greatest interest at stake, institutions serve those vested interests rather than the goals of an efficient economy. Path dependency in institutions occurs partly because those individuals and firms empowered by institutions have an interest in perpetuating the existing system.

Path dependency in technology is of interest in this volume because technology choice is central to sustainable transport. As discussed in Chapter 3, the immediate challenge for sustainable transport is to switch to less harmful transport modes within the range of existing technologies (rather than technologies, such as hydrogen, that are not mature technology systems or have doubtful environmental advantages over conventional fuels). Technological systems often exhibit various kinds of lock-in, giving rise to technological paradigms and regimes. Shared perceptions and outlooks within a technological 'community' serve to limit and focus thought and action, something that draws strongly on the momentum of existing knowledge and experiences and so contributes to incremental (rather than radical) change. Another set of explanations deal with the idea of increasing returns from existing choices, so that a cycle of marginal technological improvements leads to greater performance that in turn creates conditions that prompt the demands for further improvements.

Arthur argued that in path dependent conditions, economic allocations cannot be predicted using knowledge of the most efficient outcome or conditions; historical events 'lock-in' (inferior) outcomes. David (1985) examined the persistence and ubiquity of the QWERTY keyboard, despite allegedly superior alternatives, as an exemplar of the influence of economic path dependency, something he described as a sequence in 'which important influences upon the eventual outcome can be exerted by temporally remote events, including happenings dominated by chance elements rather than systemic forces' (1985, p. 332). He identifies the technical interrelatedness of system components, economies of scale and quasi-irreversibility of investment (such as the durability of equipment and human skills) as causal factors. Critics of David question whether the QWERTY keyboard is an inferior technology to the alternatives and suggest that there are no real incentives for change and more broadly reject claims that inferior economic options become locked-in by history (Leibowitz and Margolois 2002). According to Arthur, the initial historical events in path dependency are small scale and tend not to be noticed, but over time their influence is great. Further, these events are not inevitable and not the result of flawed decision-making, for as Leibowitz and Margolois (2002) argue, prudent and small early interventions can prevent lock-in occurring. During the early and unpredictable period when such choices are being made, the authors doubt that increasing returns explains market leadership.

These arguments on the network effect in path dependency intersect with other work on technology clusters and the benefits of co-evolution in technology.

Grübler (2003, p. 5) argues that 'whole bundles of technologies, or what we later refer to as 'technology clusters', are needed to explain the historical record of pervasive transformations within societies'. Grübler (2003) identifies historical clusters of technologies and infrastructures: 1) Textiles, turnpikes and water mills (1750–1820), 2) Steam, canals and iron (1800–70), 3) Coal, railways, steel and industrial electrification (1850–1940), and 4) Oil, roads, plastics and consumer electrification (1920–2000). Within each era there are few opportunities for other technologies to displace the dominant systems because of the strength of the linkages between the constituent parts. Notable too, in Grübler's argument, is the place of transport technologies.

Path dependency identifies that practices and techniques persist over time because the benefits of this choice accumulate and the relative costs of change are prohibitive. Over time, these arrangements become outmoded because current conditions differ from the original conditions, so that the present (and quite possibly the future) becomes a prisoner of history. Although other outcomes may have occurred, path dependency locks in practices and techniques; by implication, it suggests that a variety of different outcomes might have occurred if the conditions of initiation had differed from what occurred. Controversy in economics of whether inferior choices are made within market settings has less appeal to environmentalists, for whom the metric of environmental consequences provides a less equivocal measure of the failings of technologies, institutions and social practices. Path dependency provides an insight into why both a set of transport institutions and technologies continue to persist in the face of strong economic, social and environmental challenges.

3 Sustainable Transport

As the overview history of modern urban transport in Chapter 1 showed, mass transport, and especially motorised transport, produces benefits distributed across a spectrum of beneficiaries. There are also costs, also unequally distributed, arising from urban transport, including significant losses to environmental values and diminished opportunities for future generations. As far as the environment is concerned, these changes can only be losses, for ecology cannot ever benefit from human intervention (other than to repair or restore damages already caused by human activity). In the wake of the global environmental movement of the 1960s, a wider reckoning of motorised urban transport occurred (Minister of Transport 1963, Nader 1965, Hothersall and Salter 1977, OECD 1979). Coherent strategies and policies to address these problems only began in earnest in OECD nations in the 1970s. Policy agendas focused on three prominent concerns: pollutants from vehicle emissions, the vulnerability of oil supplies and rapid increases in fuel prices. Much of the efforts under public policy were directed towards mitigating vehicular pollution by addressing vehicle fuels and emissions (such as removing lead from petroleum) and reducing oil consumption through vehicle fuel efficiency standards and guidelines. Land use and transport planning also moved, in this era, towards recognising its role in these problems and the limitations of private vehicle use and car-based models of urban design (Banister 1994).

Contemporary overviews of the environmental costs of transport are concerned with a far wider array of issues and concerns than in the 1970s. Transport's environmental impacts include air pollution from a range of compounds (including carbon monoxide, fine particulate materials, volatile organic compounds, hydrocarbons and nitrogen oxides); water pollution from spillage and street runoff; noise pollution; land-use change from road and associated infrastructure; loss of urban amenity; resource consumption in manufacturing; operating, maintaining, and disposing of motor vehicles; and being a major source of greenhouse gases (Banister 1998, 2005, Chapman 2007, Coffin 2007). Other consequences of motorised transport systems include the inequitable social distribution of, and access to, transport services within and between nations, severe health consequences of sedentary lifestyles, the diminution of children's independent mobility, and the associated urban designs forms and structures of dispersed cities that reinforce and exacerbate many of the economic, social and environmental costs of motorised transport (Newman and Kenworthy 1999, Banister 2005, Schiller *et al.* 2010, UN-Habitat 2013, Vasconcellos 2013). An important aspect of this wider perspective is recognition of the circumstances of developing nations and the particular environmental and social costs of transport occurring there (Dimitriou 1990, Dimitriou and Gakenheimer 2011, UN-Habitat 2013, Vasconcellos 2013). Part of this enlarged understanding of these costs is bound up with considering urban transport as an entity where problems can be inter-related. As such, urban transport systems can be considered in terms of their sustainability, hence the rise of the concept of sustainable transport.

Coming to Grips with Sustainable Transport

If authors writing about sustainable transport agree on one thing, and it is a point that many express explicitly, it is that there is no agreed meaning of 'sustainable transport' (Litman 2005, Tumlin 2011, Geerlings *et al.* 2012). As environmental sustainability and sustainable development have been given such wide interpretations, ambiguity over sustainable transport is to be expected. Furthermore, the typology of sustainable transport is varied, with the concept being applied normatively (i.e. a condition that is aspired to), as a strategy, a form of analysis, a discourse of environmental politics or a process of change. Debates over sustainability can sometimes be no more than a failing to find common ground over these differences. More than a few works on sustainable transport resolve these difficulties by simply not offering any definition of their subject matter.

As definitions involving sustainability are usually open to broad interpretation, especially over key terms, there is a tendency to exaggerate the extent of these differences and overlook consistent themes. Despite the many differing definitions of sustainable transport, most share common themes, outlooks and language. Even a small sample of a vast offering conveys something of this commonality.

> Sustainable transport aims at promoting better and healthier was of meeting individual community needs while reducing the social and environmental impacts of current mobility practices.
>
> (Schiller *et al.* 2010, p. xxi):

> Sustainability refers to a development that enhances the human and natural environments now and over the long term. In this perspective, a sustainable transport system would make a positive contribution to the environmental, social, and economic sustainability of the communities they serve.
>
> (Geerlings *et al.* 2012, p. 4):

> The ability to meet society's desires and needs to move freely, gain access, communicate, trade and establish relationships without sacrificing other essential human or ecological values, today or in the future.
>
> World Business Council on Sustainable Development (WBCSD 2004, p. 5)

> The provision of services and infrastructure for the mobility of people and goods is needed for economic and social development and improved quality of life and competitiveness. These services and transport infrastructure provide secure, reliable, economical, efficient, equitable and affordable access to all, while mitigating the negative impacts on health and the environment locally and globally, in the short, medium and long term without compromising the development of future generations.
>
> (A formal definition under the 2011 Bogota Declaration on Sustainable Transport Objectives from a coalition of Latin American transport agencies)

Common to nearly all definitions and descriptions of sustainable transport, and something represented in these preceding quotes in different ways, is perhaps the most persistent idea of sustainable development, namely that it entails development that satisfies economic, social and environmental goals and in mutually supporting ways. Further, each of these pillars will be in a condition of sustainability, implying that essential and desirable aspects of these themes will be ongoing indefinitely. Recognising the needs of future human generations is a defining feature of sustainable development and occurs frequently in definitions of sustainable transport. Such is the positive character of sustainable transport that few have openly attacked the concept or its ambitions, save for a number of free-market advocates who support only market-based values (Dunn 1998).

Sustainable transport's ambiguous meaning and that of its key concepts (such as environment, society and the future), has its benefits. Such is the openness to the concept that, when combined with its positive aspirations, it has been embraced and endorsed by a wide variety of stakeholders, from environmental NGOs, governments, developed and developing nations, and international bodies to corporations (OECD 2002, ECMT 2003, European Commission 2009, WBCSD 2009, World Economic Forum 2012, SLoCaT 2013, UN-Habitat 2013). Nearly every major jurisdiction for the cities of the developed world has policies and programmes on sustainable transport, albeit varying considerably in priorities, depth and sophistication. A central element to this consensus between those with greatly differing interests and outlooks is not only a subjective interpretation of sustainability but also a shared understanding and interpretation of objective information concerning the failings of the conventional model of urban transport. As Banister (2005) and others have demonstrated, it is the view that existing urban transport systems are not sustainable economically, socially or in environmental or resource terms that enables the whole enterprise of sustainable transport to proceed.

Two sets of empirical observations underlie this conclusion. First, that transport (especially urban transport) is responsible for environmental degradation and that this causes a range of negative economic, social and ecological consequences. Second, that the trends of these effects and impacts continue to worsen at the global level, as promulgated by processes that include globalisation, urbanisation and industrialisation. Responding to these problems in ways that lessen any of these individual effects, therefore, can be deemed to constitute a positive contribution by lessening the detrimental aspects of transport. A variety of activities, therefore, can lay claim to be identified as sustainable transport. Theoretical ambiguity and practical actions on specific problems based on this rationale might be considered as contrasting conditions in sustainable transport, but critically, neither usually impedes the other from the perspectives of public authorities and corporations.

These features go some way to explaining the success of sustainable transport as a widespread set of practices. Offering development to satisfy all stakeholders makes this form of, and approach to, sustainability inherently suited to building consensus across a range of stakeholders with different interests and values. Agreement amongst civil society, governments and corporations is possible when there is a premise that a course of action will benefit all, which is the essential promise of sustainable development. Despite the broad engagement in sustainable transport plans and activities, universal consensus is rare. Persistent themes in these objections and criticisms include questioning the validity of timeframes used (such as measures only providing short-term relief), side-effects problems (such as unintended or overlooked costs which offset or outweigh the claimed benefits), failures to recognise wider consequences and reinforcing undesirable behaviours instead of producing social change. In addition to objectives over plans and activities are concerns over the legitimacy of some stakeholders' involvement (especially when they are identified as major agents of environmental and social costs). These and other concerns over what properly constitutes sustainable transport have their roots in the fundamental premises and design of sustainable development.

Sustainable Transport and the Sustainable Development Paradigm

Nearly all of the governmental, multinational and corporate activities under the banner of sustainable transport have a foundation in the concept of sustainable development, as does much of the work of NGOs in this area. Oftentimes, these efforts explicitly acknowledge the guidance of sustainable development and equate this with environmental protection. It follows that the effectiveness of these efforts will be at least partially determined by the efficacy of sustainable development in delivering environmental protection.

Sustainable development has quite specific roots in the international dialogue of the early- and mid-1980s that drew together two crises: that of the plight of developing nations and that of environmental degradation (Robinson 2004). A series of international meetings and events culminated in a set of diagnoses, proscriptions, programmes and agreements based on the argument that underdevelopment and environmental degradation were causally linked and mutually reinforcing in a vicious cycle (WCED 1987). Constructed in this way, a central idea of sustainable development is ensuring economic growth as a means to address environmental degradation and the corollary that lessening environmental degradation contributes to increasing prosperity and economic development. Lifting the boats of the world's poor was a matter of creating an economic tide of sufficient magnitude. This outlook was applied to developed and developing nations. Built into this vision is a theme from an earlier phase of environmental thinking: sustainable development considers development into the longer term by explicitly recognising the interests of future (human) generations (as opposed to the discounting applied to future values under conventional economics).

Historically, the creation of the economic, social and environmental pillars of sustainability arises from this particular formulation of the problems of global poverty, access to natural resources and services, and ecological destruction into a problem of economic underdevelopment and inadequate investment in environmental protection. That sustainable development entailed finding win-win-win solutions for the three dimensions of sustainability found wide appeal, especially in international and national policy arenas (Byrne and Glover 2002). Not surprisingly, this led to somewhat idealised and philosophical descriptions of each dimension of sustainability. Setting specific goals for the individual dimensions proved highly problematic, notably in the social and economic components (Boschmann and Kwan 2008). Translating such ambitions into policies, programmes and activities by operational areas of government, corporations and civil society proved to be only possible by limiting the ambitions for success in all three dimensions.

Critiques of Sustainable Transport as Sustainable Development

Sustainable development has attracted a range of criticisms in addition to those concerned with its ambiguous identity and themes open to broad interpretation, criticisms that are also relevant to sustainable transport as it is commonly understood and applied. These concerns deal not with what sustainable development might mean, but with its known characteristics. Underlying these critiques of sustainable development is a view that environmental protection is a primary directive, an outlook that both defines environmental sustainability and distinguishes it from sustainable development.

Environmental sustainability draws on the lessons of ecology to establish an overarching position on environmental protection and natural resource use. Setting long-term rates of harvesting for sustainable yields of environmental goods and services, which come from renewable resources, requires ensuring that rates of use do not exceed those of replenishment, rates of harvesting non-renewable resources do not exceed those at which substitutes can be developed for use and rates of waste generation do not exceed the limits of receiving environments without longer-term harm (see Daly 1977). Invariably, such ecologically based restrictions on resource use and waste generation impose limits on economic activity. Limiting economic activity does not fit comfortably with sustainable development. While sustainable development offers an escape from a paradigm that holds economic development and environmental protection as hopelessly opposed to a paradigm that seeks mutually advantageous solutions, it does so from an anthropocentric perspective and one that accepts 'Nature's' exploitation as essential for fulfilling human prospects. Further, this form of development that simultaneously embraces the interests of rich and poor nations alike is one that avoids the politically explosive issue of environmental protection necessitating a re-distribution of capital and opportunities from developed nations to developing nations. Environmental protection cannot be optimised under sustainable development; it must be necessarily traded-off against the needs of development.

Individual sustainable transport policies have featured a range of limits on resource consumption and on pollutants, such as applied to vehicle fuel efficiency standards and vehicle emissions. Strictly speaking, a number of measures under the banner of sustainable transport that involve setting limits are not, in fact, instances of sustainable development as they do not purport to also simultaneously achieve economic and social benefits (more properly, they are examples of ecological modernisation) (see Hajer 1995). Such limits often belie an underlying ethos in sustainable transport that supports economic growth. In other words, while individual policies may impose limits on economic activities, overall, the greater interests of the economy and economic growth are protected. Environmentalist concerns over economic growth are one of environmentalism's foundations, as promulgated in the seminal works of the 'survivalist' era of the 1960s and 1970s and the limits to growth thesis (Meadows *et al.* 1972). While sustainable development seeks economic growth as a means to alleviate environmental problems, such a rationale is contradictory to environmentalists, as the late Herman Daly (Daly 1990) reasoned: sustainable growth on a finite planet is a 'impossibility theorem'. Approaching the environmental and social problems created by modern transport with solutions based on sustainable development principles, is therefore self-contradictory and ultimately, largely self-defeating.

Global Urbanisation

Of all the ecological changes in the industrial era, one of the greatest has been the first, and now second, demographic transitions. This transition first occurred in the developed nations of Europe and North America between 1750–1950 and is currently underway in developing nations (UNPF 2007). As a result of global population growth, we live in a world with over seven billion people, an increase of over three billion since 1970; currently, about 82 million people are added annually to this total (UN 2014). Total population has yet to stabilise; under the UN 'medium' forecast, global population will be 9.6 billion in 2050 (UN 2014). Most of the increase to 2050 will be concentrated in relatively few nations, with nine providing over one-half of the increase and India becoming the world's most populous nation by 2028 (UN 2014). Changing demographic profiles are a feature of these demographic transitions, so that developed and developing nations have contrasting population issues, with large young populations coming to maturity in developing nations and aging populations in developed nations.

Urbanisation has many dimensions, but a major characteristic is the simple expansion of the extent of national populations living in cities, a process fostered by population growth. Large cities existed in classical and pre-modern times, but accounted for relatively small proportions of total populations; a hallmark of the rise of industrial economies is their close association with urbanisation. A century ago, only 10 per cent of the global population was urban, but urbanisation passed the 50 per cent threshold in 2008 and the process continues (UN 2008). Around one-third of the world's urban dwellers live in slums, with the highest percentage in sub-Saharan Africa (72 per cent), followed by South Asia (56 per cent) (UNPF 2007). Both population growth and rural-to-urban migration increase urban populations.

Continuing global urbanisation and population growth means a greater number of cities and greater populations in existing cities, especially in developing nations. There will be more and larger urban conurbations as existing cities merge and more so-called mega-regions linked by transport, economic relations and other connections. Some 54 per cent of the world's population was urban in 2014; this will become 66 per cent by 2050 (UNDESA 2014). Urbanisation levels vary greatly; in 2014, 16 nations (all in Africa and Asia) were less than 20 per cent urbanised, while 59 nations were over 80 per cent urbanised (UNDESA 2014). Most of the future population expansion (almost 90 per cent) – some 2.5 billion people by 2050 – will be in African and Asian cities (UNDESA 2014). Urban populations in Africa and Asia will double between 2000–30 (UN 2008). Merging of separate urban centres creates large urban conurbations, the mega-cities (i.e. cities with populations exceeding 10 million), and this process is forecast to increase the size of existing mega-cities and form new mega-cities. Currently, the largest of the world's largest 28 mega-cities are, in descending order, Tokyo, Delhi and Shanghai, with Mexico, Mumbai and São Paulo equally sized (UNDESA 2014). A majority of mega-cities and large cities are located in developing nations. There will be 41 mega-cities by 2030 (UNDESA 2014). All categories of city-sizes, from those with populations under 500,000, to the medium- and large-sized cities, will all increase in number in the coming decades.

Transport has a complicated relationship with urbanisation and with economic development. Under conventional economic reasoning, low transport costs promote both urbanisation and economic growth, providing firms with wider access to labour sources, input materials and markets, thereby promoting productivity that in turn facilitates economic growth through economies of scale and lowered production costs. Population growth, too, is generally welcomed under conventional economics as a driver of development and productivity. Greater economic connections within cities (in which transport has a major role) foster economic agglomeration, which can furnish such benefits as increased jobs, higher wages, more amenities, public transport and opportunities for commerce and cultural expressions, but often at the costs of greater pollution, crime, traffic congestion, housing prices and increased income inequality (OECD 2015). For developing nations, however, high rates of urbanisation exceed the capacities of cities to provide basic human needs at adequate standards (food, water, sanitation and shelter) and urban services, including mobility, education, health and security. In the absence of suitable institutions and economic means, developing nations struggle to create employment and business development opportunities to match increasing needs, with the resultant growing informal economies and urban slums. Sustained and rapid urban population growth is a major contributing factor to extreme transport problems in developing world cities, where providing the necessary institutions, infrastructure and expertise is a major challenge. Aspects of these problems include high pollutant levels from vehicles, high rates of road trauma, increasing inequity in mobility access, chronic road congestion and loss of urban amenity (WBCSD 2009, SLoCaT 2013, UN-Habitat 2013, WHO 2013a). Urbanisation will be a critical factor in the future of urban transport, both as being an influence on transport opportunities, demands and choices and as something whose success will be in part determined by the distribution of, and access to, urban mobility services.

Growth in Global Vehicle Production, Ownership and Use

Whoever thinks the world has entered a post-industrial era might do well to consider motorised transport; we already have over one billion motor vehicles (covering cars, trucks, buses and powered two- and three-wheelers) and on current trends are heading for twice this number (Sperling and Gordon 2009). Motor vehicles can be seen as the signature technology of the industrial era; they constitute possessions of high value and desirability, are the product of a great amount of natural resources and a colossal enterprise of manufacturing, required the output of a great research and development effort, are a major determinant of the global energy trade and a cultural fetish of great influence. Motorcars have been a transformative technology culturally, economically, politically and, of course, environmentally.

Mobility is highly motorised in developed nations and economic development is fostering increasing motorisation in developing nations. Worldwide, there are around 1.6 billion registered road vehicles, distributed as follows: high-income nations – 47 per cent, middle-income nations – 52 per cent and in low-income nations – 1 per cent (WHO 2013a). Motorisation is increasing rapidly with a 15 per cent growth since 2007, notably in middle-income nations (WHO 2013a). Sales of light-duty vehicles in non-OECD nations are now almost equal to those in the OECD and will exceed the OECD in the coming years (SLoCaT 2013). This growth has great economic, social and environmental implications. As Sperling and Gordon wrote in *Two Billion Cars* (2009, p. 1): 'Can the planet sustain two billion cars? Not as we now know them.'

Much of the motorisation of developing nations is not through cars, but with powered two-wheelers. Particularly in East and Southeast Asia, powered two-wheelers make up the majority of powered road vehicles (covering petroleum and electric motorcycles, scooters, mopeds and powered bicycles) (Posada *et al.* 2011). Taking powered two- and three-wheelers into account with motorcars, as UN-Habitat (2013) notes, the extent of motorisation in many developing Asian nations rivals that of developed nations. Despite such levels of motorisation, per capita greenhouse gas emissions from the transport sectors in developing nations is far less than those of developed nations, although motorised transport is a major contributor to air pollution in most large cities in developing nations.

Recent growth in electric-powered bicycles (e-bikes), particularly in the case of China, has been exceptional. Between 1998–2005, annual Chinese e-bike sales grew from 40,000 to 10 million (Weinert *et al.* 2007). Estimates for annual e-bike sales vary considerably; in broad numbers, current global annual sales are around 40 million, with around 30 million in China, around 2 million in Europe, 450,000 in Japan and perhaps 200,000 in the US. City government policy in China has played a role in this trend because e-bikes produce no tailpipe emissions, provide high energy-efficiency mobility and contribute less to road congestion than motorcars. Given the problems created by the proliferation of small capacity motorcycles across Asia in recent decades, the appeal of e-bikes to city authorities is obvious (noting that several cities in China have banned petrol-powered two-wheelers). On the question of the place of e-bikes in China's evolution in transport, Cherry and Cervero (2005) considered that e-bikes were a lower-cost high mobility form of transport in their own right rather than the transition between public transport and motorcar ownership. Advocates consider e-bikes to have the potential to displace the motorcar for local trips on a large scale, although to date China is the only mass market worldwide. E-bikes, however, have higher environmental costs than bicycles, although the pattern of trips differs between the modes (such as higher trip distances for e-bikes). China's experience has shown an increase in e-bike fatalities and injuries, raising safety concerns over e-bikes in mixed road traffic. As to a complete reckoning of the potential environmental and social benefits of e-bikes, the jury is still out (Rose 2012). Future growth of the e-bike market remains somewhat uncertain, although considerable expansion of the market is generally expected (e.g. Weinert *et al.* 2008).

Including all types of motorised transport does not alter the great differences in motor vehicle ownership between the world's nations. Such are these differences that the global average ownership rate of around 148 vehicles (all types) per 1,000 people is not at all indicative of national circumstances. In 2012, at the low end of the scale of vehicle ownership (as vehicles per 1,000 population), are India: 24, Africa: 34, Indonesia: 72, Far East Asia: 77 and China: 82 (Davis *et al.* 2014). Nations with higher national wealth show higher ownership rates, such as Middle East Asia: 124, Brazil: 187 and Central and South America: 177 (Davis *et al.* 2014). Highest ownership levels are found in the wealthiest nations in East Europe: 322, West Europe: 589 and the Pacific nations: 569 (Davis *et al.* 2014). Heading the list of ownership levels is the US with 808 vehicles in 2012, some 1.19 vehicles per licenced driver and 2.08 vehicles per household; 19.3 per cent of households have 3 or more cars (Davis *et al.* 2014). Interestingly, this US rate is down from the high point of 845 in 2007, a level of car ownership that may never be surpassed for a large nation. Certainly, the historical US dominance of global car ownership will not be repeated, for in 1960, 46 per cent of the world's registered cars were in the US (today, that figure is 16 per cent) (Davis *et al.* 2014).

Global vehicle fleets continue to grow, with global car production increasing by 34 per cent between 2000–12; car registration has followed suit, increasing by 2.5 per cent annually since 1990 (Davis *et al.* 2014). Broadly, the international picture is that of high vehicle and slightly declining ownership in richer nations and low vehicle ownership in developing nations, with very rapid rates of increasing car ownership in middle-income nations. China, the world's most populous nation, has a relatively low per capita vehicle ownership but it has been increasing at a high rate (some 16 per cent annually since 1990), so that now it has 52 million registered cars, up from 3 million in the year 2000 (Davis *et al.* 2014). India, the second-most populous nation, now has some 18 million cars and a fleet growth of 10 per cent annually over the period 1990–2012 (Davis *et al.* 2014). By comparison, the US has the most vehicles, with 232 million cars and light trucks in 2012 (Davis *et al.* 2014). By way of historical comparison, US ownership was similar to contemporary India in 1915, to China in 1920, to Central and South America in 1925 and to Eastern Europe in 1950 (Davis *et al.* 2014).

Centres of global motor vehicle production and the strength of the emerging markets are linked. While the established large manufacturers in Europe, Japan and North America have struggled economically in recent decades, with relatively stagnant domestic markets, the newly industrialising and more recently industrialised nations with growing demands are also home to new manufacturing businesses. China is the most prominent of these new vehicle producers; its car production increased by a spectacular 1,670 per cent between 2000–12 (Davis *et al.* 2014). Of the 55 million vehicles manufactured in 2012, almost 11 million of these were of Chinese origin (Davis *et al.* 2014), making it now the world's leading manufacturer (followed by Japan and Germany, respectively) and primarily serving a domestic market. Although its industry receives relatively little attention in the OECD, India's car production increased three-fold between 2000–12.

Critically, a key environmental issue in motor vehicle consumption is the size of the future vehicle fleet and whether or not the historical association between rising private vehicle ownership and rising national incomes will continue into the future (World Bank 2010). It is generally taken that US$5,000 average annual per capita income is the threshold beyond which rapid motorisation (as car and truck ownership) occurs, an effect that tapers off somewhere around the US$35,000 income level. Clearly, the greater the number of motor vehicles, the greater the total environmental and social impacts, but also as private car ownership rises, there tends to be increased dependency on private vehicles for mobility and diminished public support for public transport. Car ownership and use are also positively linked, such as measured by distance travelled per capita by motor vehicles. There are exceptions to these associations, but the US exemplifies this concern, for in the nation with the highest car ownership, in 2012 only 5 per cent of commuters used public transport, 3 per cent walked and less than 1 per cent cycled (Davis *et al.* 2014). In developing nations, as noted above, public transport systems are often not of high quality or in large supply, making private motorisation attractive as a means of mobility.

Interestingly, there are some trends in place that bring into question future rates of motorisation. Firstly, there may be a negative association between motorisation and urbanisation; i.e. when populations become highly urbanised in developed nations, motorisation declines. This appears to have been the case for 'mature' cities such as London, Paris, Stockholm and Tokyo. Many different causative factors have been suggested, including the effects of road congestion, rising costs of car ownership, difficulties with parking, environmental awareness and substitutes for personal mobility, such as modern communication technologies. Secondly, public policy can also be an important influence, such as providing incentives for reducing car use through increased public and active transport investments. Disincentives from public policy can also be effective, such as high taxes on car ownership, congestion and road access charging and parking controls. These same factors may also be responsible for the phenomenon of reducing driving in developed nations. Thirdly, as discussed below, there are several environmental (notably responses to limit future global climate change) and resource restraints (prominently, oil as a non-renewable resource) likely to limit future levels of motorisation.

Changing patterns of car use requires more than providing public and active transport infrastructure and services as alternative mobility options, although these are essential. Efforts by policymakers to change the behaviour underpinning the use of unsustainable transport modes have to tackle the perceived attractive attributes of private motor vehicle use, notably speed, convenience, comfort and freedom. Motorists' attitudes have proved resistant to change from many conventional policy approaches, prompting investigations into possible psychological explanations. This research has identified phenomena such as social norms, habitual behaviour, individual perceptions and personal identity as cutting across such conventional explanatory factors such as socio-economic status. Anable (2005) contributed to these insights with her classification of the attitudes of those who could potentially choose an alternative means of transport to the motorcar, classified on a continuum of receptiveness to switching: Die hard drivers, complacent car addicts, malcontented motorists, aspiring environmentalists, reluctant riders and car-less crusaders. These sorts of findings suggest that not only is there a need for 'soft' measures (e.g. regulation, taxation and personal travel plans) in addition to the usual 'hard' measures (e.g. infrastructure), but that behaviour can have different causes and that similar attitudes result in different behaviour, thereby requiring a suite of policies to suit different attitudes and behaviours (Cairns *et al.* 2004, Anable 2005).

Road Transport and Greenhouse Gas Emissions

Global climate change resulting from the anthropogenic production of greenhouse gases is a major environmental crisis, the seriousness of which now requires no further elaboration. Globally, transport is a major source of greenhouse gas emissions, producing 7,187 million tonnes of carbon dioxide (the major greenhouse gas) and accounting for 23 per cent of global carbon dioxide emissions in 2012 (IEA 2014). Three-quarters of these transport emissions are from road transport (IEA 2014). A critical factor in transport emissions is the transport sector's reliance on a single source of energy; 94 per cent of its 2010 demand was met through oil use (with biofuels, electricity and natural gas making up most of the balance) (IPCC 2014a). Light duty vehicles consume about one-half of all transport energy (IPCC 2014a).

Transport consumption is highly skewed; according to the IPCC, only 10 per cent of the global population accounted for 80 per cent of all motorised passenger kilometres, mostly within OECD nations (IPCC 2014a). Greenhouse gas emissions from transport are also skewed, with transport's share of national emissions ranging from 3–30 per cent (IPCC 2014a). Despite the concentration of motorised mobility in wealthier nations, developing nations have been closing the gap rapidly. Carbon dioxide emissions from non-OECD nations have now reached 80 per cent of those of the OECD (IEA 2014).

Growth has been most rapid across Asia, as associated with rising national GDPs, while there are signs of weakening growth and lower light-duty vehicle use in some OECD nations (IPCC 2014a). Income levels and emissions from transport are positively linked, most noticeably at higher income levels, but this varies greatly within and between nations and cities. It follows that there remains considerable potential for further growth in transport in developing nations, although it may not rival the extent of private motorised travel in developed nations due to factors such as the constraints on infrastructure and higher urban densities (and the future availability of relatively low-cost transport energy supplies).

Despite the initial global agreement for nations to limit greenhouse gas emissions in 1996 (under the UN Kyoto Protocol), transport emissions continue to increase. Greenhouse gas emissions from transport have doubled since 1970 (at 2.8 $GtCO_2$-e), are the fastest-growing end-use sector and reached 7.0 $GtCO_2$-e in 2010, with 80 per cent from road transport (IPCC 2014a). Some 27 per cent of global energy consumption is by transport, of which 40 per cent is urban transport (IPCC 2014a). At the national level for richer nations, the profile is similar. In the US, transport accounted for 28 per cent of that nation's 2012 greenhouse gas emissions (Davis *et al.* 2014). In developing nations, and especially the least developed nations, increasing motorised mobility is sought as a means to increase health, education and economic opportunity, a trend at odds with limiting transport's greenhouse gas emissions (IPCC 2014a).

Mitigating the combustion of fossil fuels remains the major requirement for limiting the extent of future global climatic warming. Ongoing diplomatic activities under the UN Framework Convention on Climate Change (UNFCCC) seek to set international and national targets and timetables for greenhouse gas emissions reduction. Under the UN Copenhagen Accord, the international community agreed to limit global warming during this century at 2 degrees Celsius above the current global average temperature, this being deemed that to prevent 'dangerous interference' with the global climate system (UNFCCC 2009). Efforts to limit rising greenhouse gas emissions under the UNFCCC have been ineffectual to date and there is a gap between current national pledges for mitigation and the emission cuts required to meet the 2 degrees warming target (UNEP 2014a). Climate scientist, James Hansen, is amongst those who consider 2 degrees of warming to have catastrophic consequences and considers that warming should be limited to that which has already occurred (about 1 degree Celsius or so), with atmospheric carbon dioxide levels drawn back down to 350 ppmv (these currently exceed 400 ppmv) (Hansen 2009). Even meeting the 2-degree target requires massive cuts in greenhouse gas emissions; it would require cuts of 49–63 per cent below

2010 emissions by 2050 according to the UNEP (UNEP 2014a) and 40–70 per cent according to the IPCC (IPCC 2014b).

Emission reduction targets in climate change policies usually apply to the whole economy within the jurisdiction of a particular authority (be it national, state, city or other jurisdiction), although it is not unusual for some favoured segments of the economy to be exempted for either political or pragmatic reasons. Individual economic sectors, such as transport, typically are not allocated a specific emissions reduction target. National targets for the short-term have been generally modest to date; under the (first phase of the) Kyoto Protocol, the average reduction of emissions from the (usually 1990) baseline to the period 2008–12 was around 5 per cent. Longer-term targets have been more ambitious; the EU aims to cut greenhouse gases by 40 per cent by 2030 and 80–95 per cent by 2050 (European Commission 2011). In most jurisdictions, the transport sector has lagged behind other economic sectors in reducing emissions; for most nations, transport emissions have been increasing over the period covered by the Kyoto Protocol (IPCC 2014a). More ambitious targets, such as that of the EU, have profound implications for the transport sector, as 'de-carbonisation' has proved very difficult to date at scales above that of the household and small firm.

Safety and Health Impacts of Road Transport

Mobility through road transport incurs a high toll in road trauma. According to the WHO, there are 1.24 million lives lost annually on the world's roads and an additional 20–50 million people injured (giving a global fatality rate of 18 per 100,000 of population) (WHO 2013a). Sharma (2008) argues that road traffic injuries constitute a major global health crisis that disproportionally burdens the poorest nations, with vulnerable road users neglected by most policy interventions. Young adults account for the majority of trauma victims, making it the leading cause of death for those aged 15–29 years (WHO 2013a). Vulnerable road users – pedestrians, cyclists and motorcyclists – are over-represented in road trauma, accounting for one-half of road fatalities (WHO 2013a).

Although the low- and middle-income nations have only one-half of the world's road vehicles, they have 91 per cent of road fatalities (WHO 2013a). In poorer nations, the loss or injury of household members greatly reduces household income, productivity, savings and wider prospects for prosperity. One estimate put total global costs of road trauma at US$518 billion (WHO 2013a). Road trauma can be reduced and successful public policy interventions have lowered road trauma where introduced (addressing the risk factors of speed, drink-driving, helmets, seat belts and child restraints). These efforts have not been widespread: such initiatives have been used comprehensively in 28 of 195 nations surveyed, covering 7 per cent of the world's population (WHO 2013a). Such measures may offer greater risk reduction for motor vehicle users, but it is the vulnerable road users in poorer nations who are in most need of protection (Ameratunga *et al.* 2006).

As terrible as global road trauma is, its impact on human health is probably less than that caused by air pollution. In 2013, WHO's International Agency for

Research on Cancer classified outdoor air pollution as carcinogenic to humans (WHO 2013b). WHO estimated the global death toll from air pollution in 2012 at around 7 million, elevating it to what it describes as the world's largest single environmental health risk (UNEP 2014b). These estimates are higher than previously reckoned, partly as a result of improved mortality data and new evidence of health risks and air pollution. Also, these findings combine indoor and outdoor air pollution (with 3.7 million of the aforementioned deaths attributable to outdoor pollution), so that the transport pollution component is but one of several pollution sources. Whilst the air quality in richer nations has improved, the opposite is true of middle-income nations; the majority of deaths from this occur in middle- and low-income nations. WHO report that road transport accounted for 50 per cent of the 2010 health impact costs (of both mortality and morbidity), some US$1.7 trillion in the OECD nations (including China and India); the authors note that there is insufficient evidence to estimate these costs for all developing nations (OECD 2014). Europe, which we often regard as a world leader in environmental reform, is far from being free of such problems; one-third of city-dwelling Europeans are exposed to air quality that exceeds the EU standards (EEA 2013). Nearly all city dwellers were exposed to fine particulates (as $PM_{2.5}$) and ozone levels above WHO guidelines. Road transport is a major contributor to European air pollution, responsible for 32 per cent of NOx.

Peak Oil

While the prominent transport energy issue on the 1970s, namely the oil crises of 1973 and 1979, resulted from deliberate manipulation by OPEC and outcomes of the export disruptions due to the Iranian Revolution, peak oil is the looming oil supply crisis for the next generation (Roberts 2005, Deffeyes 2009). Global energy use in transport results from a staggering logistical maze of oil extraction, transfer, refining and retailing. Even the volumes involved are impressive, with global crude oil production of 76 million barrels a day in 2013, petroleum production and consumption (including natural gas plant liquids) at 88 mbd; the largest national consumer was the US with a 21 per cent share (at 18.6 mbd) (Davis *et al.* 2014). As peak oil expert, Kjell Aleklett, expresses it we live in a 'world addicted to oil' (Aleklett 2012: 17).

Briefly put, peak oil is the point of maximal production followed by an inexorable decline, understanding that oil is a non-renewable physical resource. Global oil consumption has exceeded the rate of discovery of new resources since the early 1980s (Aleklett *et al.* 2010). In all likelihood, the world has passed the point of maximal conventional oil production, so that the remaining reserves are less than the oil produced to date (Kerr 2011). Rather than the world running out of oil from conventional sources, the key problem of peak oil is that the cost of oil will rise as reserves draw down (and demand goes up as the world's vehicle fleet expands).

Given the extent to which the motorisation of urban mobility has been built on cheap oil, the cessation of low prices proffers significant social and economic change, both nationally and globally. Opinions differ sharply on the scale and scope of these effects: optimists see technological development outpacing any deleterious effects of reducing supply of conventional oil (Clarke 2007). With the urgency of known and pressing environmental crises (especially global climate change), taking an optimistic view of continued high energy use is necessarily indifferent towards environmental sustainability. Pessimistic speculations, which are numerously published, paint a doomsday scenario of resource wars and massive social breakdown (Kunstler 2005, Dyer 2008), reminiscent of the environmentalist survivalist school of the 1970s. Many of the basic contours of the business-as-usual case of peak oil are less controversial, but no less alarming in their implications, especially for poorer nations and the economically disadvantaged (Roberts 2005). Peak oil also provides an opportunity for social change, prompting many authors to recognise the potential for progressive reform towards environmental goals for communities (Murphy 2013), reformers for a sustainable economy (Lovins 2011), city planners (Newman *et al.* 2009) and transport planners (Gilbert and Perl 2010).

Certainly, the lessons of the 1970s oil crises were economically harsh, with many holding the 1979 oil crisis responsible for a world economic recession. Economically, a substantial rise in transport energy costs as caused by peak oil will depress productivity. Significantly higher fuel costs may curb future growth of the world's vehicle fleet, acting to restrain the uptake of privately owned motor vehicles in developing nations with rising per capita incomes. Flow-on effects of higher transport costs will include higher food prices, applying to both mechanised agriculture and food freighting (Pfeiffer 2006). Overall, the costs of living in many places would invariably increase as a direct result of higher transport costs and the associated increased costs in a wide array of goods and services. Some consider the basic tenets of global trade and international passenger movements will be undermined to the extent that globalisation will be curtailed (Curtis 2009). Resource conflicts within and between nations become more likely as global energy security declines. We need also to be mindful of the geo-political implications of the world's reliance on oil for transport as the world's major reserves remain within politically unstable nations and regions (OPEC has three-quarters of the world's crude oil reserves). As oil reserves become even more valuable in nations where oil income is controlled by unelected elites, political tensions are unlikely to diminish.

Some pundits believe that the oil production from unconventional sources (such as biofuels, coal-to-liquids, tar sands, oil shales and the so-called heavy oils) can rise to help meet future oil demand as conventional oil production declines, but the viability and extent of this substitution is open to question (IEA 2013, USEIA 2013, BP 2015). Currently, there are no large-scale and immediate low-cost liquid fuel alternatives to conventional oil. What is settled in this debate is that unconventional oil production is relatively expensive and is only economically viable when conventional oil is of high cost. It is important, therefore, not to confuse a solution for addressing shortfalls in energy supply with the problem of high transport energy costs. There are other major concerns over unconventional oil sources: firstly, they tend to have a very low *energy investment* (meaning that there is not much difference between the energy required to produce the energy and the energy yielded, relative to conventional oil). Secondly, there are a number of harmful environmental consequences involved in producing non-conventional energy, as shown by the controversies over Canada's oil sands mining (Rooney *et al.* 2012).

Following the oil crises of the 1970s, the environmental sustainability discourse was dominated by the themes of the profligate use and waste of this valued and diminishing resource, with the 'limits to growth' thesis giving way to the survivalist school of environmentalism. Threats to the security of energy supply were sufficient to attract mainstream attention and widespread public policy responses. Today's version of this debate is peak oil, but in many senses, this has become a redundant argument. Given the environmental and social costs of oil combustion, notably the consequences of global climate change, the key environmental challenge is not the end of cheap oil but that of ensuring that the remaining fossil fuels remain in their geological stores. Associated with this change in the debate is a shift from an anthropocentric concern that transport will become more expensive and constrained by energy availability to one over the natural and social impacts of climate change and other negative effects arising from the transport sector's consumption of fossil fuels.

Mobility Access and Inequality

There is great social inequity in access to, and use of, transport services. Although this is a blanket statement, it is a strong generalisation that the opportunities for access to mobility are uneven within societies according to geographic, economic, social, political and individual characteristics. Transport poverty is most apparent in poorer nations and in lower-income groups and it contributes to household poverty. Opportunities for travel are linked to income levels, with the poor having fewer choices, so that in developing countries many people have no other travel choice other than to walk. Public transport can provide universal access (as it does in wealthier European nations), but where these services are inadequate, the wealthier use private cars and the poor walk and cycle, and as their income increases often graduate to motorcycles, then use taxis and shift to cheaper cars. These are also the circumstances that promote informal transport in developing nations. In the international context, the problem is the asymmetry in mobility access between those with a dependence on the unsustainable private motorcar and those with marginal mobility options of any sort.

Access to mobility, although a commonplace phase, is a difficult concept to pin down, partly because mobility itself is so broad and accessibility concerns both the features of a transport system and the characteristics of individual people (evoking the idea of *motility*). Central to issues of access to mobility are service availability, service affordability and acceptability of the service to the traveller. Furthermore, access and equity can never be reduced to simple indicators without encountering subjectivity and political judgements. Egregious examples of inequitable mobility access, however, express core concerns of mobility y enough. UN-Habitat (2013), for example, reports that 20 per cent Mexico City spend more than 3 hours commuting and that the costs nformal transport can consume 20–25 per cent of daily wages in such as Buenos Aires, Delhi and Manila. A major initiative under the UK's Blair government on accessibility revealed extensive mobility disadvantage in that relatively prosperous nation. For example, over 12 months, 1.4 million people missed medical appointments due to transport problems, 38 per cent of those seeking work found a lack of transport a barrier to getting work and children in the lowest socioeconomic group were five times as likely to die in a road event (Social Exclusion Unit 2003). Overall, the study reported on the difficulties of those without access to a car or only poor public transport services had getting access to work, healthcare, education, food shopping and other services.

Failings in such access can arise through a range of deficiencies in transport systems, including high costs of services, an absence of services (and return services) at times and locations when required, an absence of access to desired destinations, physical and intellectual barriers to users, unsafe or hazardous conditions and the absence of connecting or integrated services. In a number of developing nations in Africa, Asia and Latin America, public transport has been in decline as the colonial era of monopolised bus companies gave way to smaller companies under public contracts, as across Latin America and to extensive informal transport in many African cities. Across China, in a number of Indian cities and in parts of Eastern Europe, public transport monopolies remain in use. Declining public transport services obviously create greater inequity in urban mobility, an inequity that has several dimensions. Transport is the means to other ends; it is the service that enables access to other services, prominently employment, shopping, health, commercial services, education, recreation and socialising. For those suffering transport disadvantage, the consequences are reduced opportunities to use and enjoy these services, so that transport disadvantage invariably results in economic and social disadvantage. It follows that the geographical and social distribution of socioeconomic disadvantage tends to match that of transport disadvantage and vice-versa for those who are better off.

In richer nations, cities have often become car-centric, leaving those on low incomes without car access relying on public and/or active transport and forcing those on lower incomes into car ownership. Residential land use patterns in these cities can reinforce mobility opportunities where land values reflect accessibility to transport infrastructure and services, so that the income poor are forced into transport poor locations. Consequently, further investment in road transport infrastructure, especially in the form of mega-projects, further promotes car use and increases mobility disadvantages (in many developing nation cities, such projects displace poor communities and can add greatly to the safety risks for local populations). A major source of inequity is often invisible, namely that arising from public subsidies to road transport and car users and the uncaptured externalities of car use that deprive those without car access.

However, the absolute dominance of the car appears to be weakening, with car ownership and use in many richer OECD nations either stabilised or declining slightly over the last 10–15 years (Metz 2013, Kuhnimhof *et al.* 2013, Van Dender and Clever 2013). Measures of total passenger kilometres travelled show this trend in the US, UK, France, Japan, Australia and elsewhere (Van Dender and Clever 2013). Explanations vary, including the after effects of the global financial crisis, higher fuel prices, slower population growth, ageing populations, increasing urbanisation and successful policy interventions. Interestingly, there is a socioeconomic aspect to these changes, with per capita car use declining for young adults (notably men). Again, there are several contending explanations, including delayed family formations, unfavourable employment conditions, greater use of public transport and the influence of the Internet on travel behaviours. It is not yet clear whether the declining car use in OECD nations is an interregnum of long-term growth, a saturation in which there is no interest in further car travel or the beginning of a trend of declining car use.

Technology to the Rescue? Limits of Green Cars

Alternative fuels and vehicles are promoted as the solution to the failings of the conventional motorcar and there has been considerable investment in the quest for the 'green car' since the 1970s. This is a field with much optimism and high expectations for technological solutions for addressing sustainability in transport (European Commission 2009, OECD/ITF 2010, World Economic Forum 2012, IEA 2005, 2012, 2015a). Optimism, in a sense, reflects a basic problem with the new road and fuel/energy source technologies, i.e. they have yet to make a significant cut in current emissions and at present, have little potential for greatly reducing future emissions (Moriarty and Honnery 2008, World Bank 2010, Kopp *et al.* 2013).

Reviews of alternative fuel vehicles typically find the most cost-effective strategy to be that of efficiency improvements to the conventional light-duty vehicle (IEA 2006, 2012, ECMT 2007), with the hybrid electric-petroleum vehicle to be the best alternative energy prospect for the immediate future (depending on the source of electricity). These alternative fuels/energy sources are of interest for addressing several issues, including energy security, rising costs of conventional fuel and environmental protection. Corporations with commercial interests in motor vehicles, transport infrastructure, transport fuels and energy sources have supported research and development into improving efficiency of internal combustion engines and their fuels and into alternative fuels and vehicles (WBCSD 2004, World Economic Forum 2012). Government agencies have also promoted the potential of alternative fuels and vehicles over a long period, notably from Europe and the US (e.g. ECMT 2007, European Commission 2009).

Finding viable alternative fuels and energy sources has proved an extremely difficult task in light of several constraints, which include the need for low environmental impacts on a life-cycle basis, a high energy density, something which offers an energy gain, is affordable and capable of being stored and supplied to hundreds of millions of vehicles. Furthermore, in light of the demands of greenhouse gas emissions reductions, the timeframe for replacing conventional vehicles

PCC 2014a). Progress in replacing the
w-carbon and alternative fuel vehicles
the 1970s that initiated much of this
nal figures, a little over one million
ive fuels and energy sources out of a
Most of these alternative fuel/energy
than 100,000 powered by electricity
lectric vehicle sales exceeding 1 per

duction target of 80 per cent from
obility would have to be achieved
s. When positing the minimisation
stainable transport, it is envisaged
be consistent with such a 2050 (or stricter) target. Technically, such a target is deemed achievable in the US using much higher efficiency vehicles, and with a large proportion of the vehicle fleet using biofuels, electricity and hydrogen (but only as produced with low greenhouse gas emissions) (NRC 2013). This proviso on producing alternative fuels/energy sources with low greenhouse gases is critical because many of the short-term leading contenders are strongly compromised in this regard.

Electrification of mobility, the current leader in alternative transport energy, is widely held to be the most likely long-term successor to conventional light-duty vehicles. While the long-term option may be battery electric vehicles, the short- and medium-term options appear to be the transition technologies of hybrid electric vehicles (i.e. petroleum and electric power) and plug-in hybrid electric vehicles. Electricity is, however, only as clean as the source of its generation and that is a major barrier to electric vehicles from an environmental perspective. Electric vehicles are only viable if electrical current is taken from renewable energy sources and largely exclude fossil fuel-powered generation, the current source of the bulk of global electricity (IEA 2014). Biofuels, in theory, could supply large-scale liquid fuel stocks for transport but face severe shortcomings. The commercial biofuels currently available, the so-called 'first generation biofuels', offer few greenhouse gas emissions savings over conventional fuels (Hill *et al.* 2006, IEA 2015b). Great doubts exist over whether such biofuels can ever be scaled up sufficiently to replace oil-based fuels without diminishing food production or encouraging forest clearance for cropping, outcomes that would be particularly iniquitous in a world still marked by starvation (Worldwatch Institute 2012). Certainly, a set of tight international controls would be needed to avoid such negative consequences (Tilman *et al.* 2009). Second generation biofuels produced synthetically from inedible cellulostic biomass could resolve these problems, but are not yet a commercial product. There have been periodic high hopes for hydrogen as an energy source, but its commercialisation for light-duty vehicles is yet to occur, with major production and storage issues yet to be resolved (Romm 2004, IEA 2015b). Few transport energy scenarios feature any meaningful contribution from hydrogen energy sources in coming decades.

While these technologies may address the problem
(putting aside the matter of whether sufficient technolog
to shift from the development to maturity stage), relying
solutions depends on two key determinants: 1) Whether the
will be applied, and 2) Whether they can be applied with suffi
the global scale, relative to the demands of climate change mitiga
that neither condition is likely to be met. Moriarty and Honnery (200
ple, take a greenhouse gas emissions reduction target of 75 per cen
from current values based on estimates on requirements to restrict global
ing to the 2 degrees Celsius, finding that: 'Assuming that percentage reduc
in car travel emissions must match overall reductions, emissions per car p-k
[passenger-kilometres] would need to fall about 14-fold by 2030 compared with their present value' (2008, p. 1719). They point out that if this reduction burden were distributed equally between nations, then developed nations face very high reductions from current levels.

Forecasts of the future alternative fuel/energy sources and low-carbon vehicle use are not particularly encouraging. As suggested above, the limit of these technology solutions is that substituting these alternative vehicles for conventional vehicles at the scale and in the timeframes required will not occur under business-as-usual trends. At least two other factors add to this dilemma: firstly, continuing growth in global vehicle sales erodes the emissions savings made through technological improvement and change (IEA 2006, OECD 2008). Secondly, despite whatever foreseeable changes are made to light-duty vehicles and their energy sources, they and their energy, infrastructure and other supporting systems will continue to have a large ecological impact (such as well-to-wheels studies of electric vehicles reveal) (see, e.g. Elgowainy *et al.* 2009). Alternative fuels and vehicles do not fundamentally alter the extent of energy use, resource consumption and waste production that light-duty motor vehicles incur in providing mobility to their passengers for every trip they take.

Policy Responses for Sustainable Transport

There are a number of aphorisms for sustainable transport strategies, some being drawn from that of resource conservation's 'reduce-reuse-recycle'. One of the most popular is 'avoid-shift-improve' which encompasses most of the strategic measures used in public, corporate and NGO policies and programmes and offers a simple categorisation of a very broad and diverse field of activity. Although each of the themes can be implemented in many different ways, there is a rough logical association between each theme and its scale of attention, with 'avoidance' being largely concerned with transport and urban systems, 'shifting' dealing largely with trip choices and 'improvements' generally focusing on vehicles and energy sources. Jurisdictions and stakeholders also differ somewhat between these strategies, so that different sorts of policy instruments and approaches are employed within these strategic themes.

Avoid

Strategies of avoidance have the basic intention of reducing or avoiding travel and are, in a sense, the most basic way of reducing social and environmental impacts. Nearly all of the effects of transport vary quantitatively with transport activity levels, so the greater the amount of mobility per transport system, the greater the resultant detrimental impacts. Finding ways to eliminate or shorten trips is, therefore, a critical component of sustainable transport; some argue that reducing travel is the essential strategy of sustainable transport, with the 'shift and improve' strategies being of secondary importance. Urban and transport planning, together with travel demand management, are the common approaches for reducing travel demand. Integrating these planning functions, an association that has enjoyed far more theoretical agreement than actual implementation, can use urban design in its broadest sense to locate popular destinations and transport infrastructure and services in ways to minimise or avoid travel. Urban spatial planning can place destinations relating to employment, health, education, commerce and other services in locations that reduce travel distances. Associated transport planning can be coordinated with such planning, ensuring that public and active transport options are available and an attractive means of mobility (and provide a viable alternative to private motor vehicle travel).

Some trips can be made redundant through the use of modern communication and information technologies, of which replacing commuting by telecommuting is perhaps the best-known exemplar. Although it perhaps too early to fully understand the influence of the Internet in this regard, there is a broad consensus amongst researchers and policymakers that it will exert a considerable influence on future urban transport; the extent to which it has and can reduce total travel time is uncertain at this time. Trip length reduction policies include urban planning guidelines, incentives/subsidies for low-carbon urban developments and mixed land-use developments. Travel demand reduction policies aim at behaviour change, using information programmes on the costs of travel, trip planning services, promoting telecommuting and implementing financial disincentives for private car use, of which parking fees are frequently used.

Shift

Strategies of shifting refer, generally, to changes in transport mode as a way to achieve the goals of sustainable transport by increasing the use of those modes with relatively low impacts and reducing the use of the least sustainable transport options. Clearly, the central goal of these strategies is to shift as many journeys down the ladder of transport modes ranked by their environmental impacts, but most obviously the key shift is from private motor vehicles to public transport. Shifting trips from motorised forms to non-motorised forms (i.e. walking and cycling) brings multiple benefits in addition to the obvious lowering of vehicle emissions and pollutants. Reduced motor vehicle use eases road congestion, provides health benefits to participants, adds to local urban amenity and can be made part of local economic endeavours. Non-motorised transport can also be linked to public transport to provide a substitute for motor vehicles for longer urban journeys.

Mass transport is the means by which most passenger trips can be undertaken with low environmental and social impacts, drawing primarily on suburban rail, light rail and buses (including bus rapid transit systems). There are also options for vehicle sharing, so that the environmental cost per passenger per vehicle is reduced. As discussed in later chapters, there are a variety of options for vehicle sharing, including carpooling, ridesharing, car share schemes and bicycle share schemes. Policies to shift travel to public and active transport and to prevent or discourage the uptake of private vehicle use include road tolls of various kinds, such as congestion charging, tollways, access charging, restrictions on road use, vehicle parking restrictions and pricing of car parking, restrictions on vehicle ownership and use and pricing of vehicle fuels (e.g. fuel taxes). Incentives for public transport use include improved and more numerous services, preferential treatment for public transport on roads (such as bus lanes, bicycle lanes and dedicated rapid bus transit lanes) and dedicated infrastructure for walking and cycling (as this encourages inter-modal trips).

Improve

Strategies of improvement seek to reduce the social and environmental costs through measures to improve the performance of individual modes of transport (including transport infrastructure), covering vehicles and transport fuels and energy sources. More than the other two themes, strategies of improvement feature technological innovation and development. Vehicle and fuel technology options include improvements to conventional internal combustion engine technologies (such as efficiency improvements, though stop–start technologies), lower-emission fuels (biodiesel) and alternative fuels and energy sources (primarily natural gas, hydrogen fuel cells and electric vehicles). Policies to improve vehicles, fuels and infrastructure include subsidies for alternative fuels, special assistance with new infrastructure for alternative fuels and energy sources, financial incentives for more fuel-efficient or low-carbon vehicles, campaigns for more fuel-efficient driving and vehicle choice, and labelling of vehicles with fuel/emissions information. Similar approaches can be used to address vehicle emissions, but additional measures include fuel efficiency standards and guidelines, eco-driving campaigns and revising speed limits downwards.

Towards Environmentally Sustainable Transport

Generalisations about sustainable transport are difficult to come by not only because of the fluidity and openness of the concept, but also because it is a blanket term that covers a constellation of environmental, social and resource problems and solutions. As a result, sustainable transport has, itself, been on a journey of evolution, marked by a broadening and deepening of the concept and its applications. Several factors lie behind these changes:

- New understandings of urban and transport systems that widened the scope of the interests of transport decision-makers and attracted other disciplines' interest in transport, such as health and social welfare.
- Recognition by researchers and policymakers that the narrow scope of vehicle-based pollution reduction measures and idealistic land use planning aspirations were insufficient to address the rising environmental and social costs of transport and that a broader, more comprehensive and more integrated approach to identifying and creating solutions was required.
- New technologies and technological applications that offered new tools and solutions to transport problems, such as those used to monitor vehicles in congestion charging systems.
- Discovery of new environmental and social problems, such as those related to the causes and potential impacts of climate change, and revelations of new effects of existing problems, such as sedentary lifestyles, obesity and the health impacts of urban air pollution.
- New constituencies and stakeholders articulated their concerns over environmental and social effects and called for governments to act, corporations to respond and who organised community-based solutions.

As the subject of international, national, state and city public policy, sustainable transport could hardly have been more successful given the extent to which it has been adopted by policymakers influencing urban transport and in initiating research and development into ways to address the environmental and social costs of urban transport. To this can be added the considerable efforts by the corporate sector, especially in motor vehicle and alternative fuels and energy sources. A considerable body of knowledge has been built up around the world as a result of these efforts, comprising a range of effective policy solutions, best practices, new technologies, urban designs and ways of achieving behavioural change across the strategies of 'avoid-shift-improve'.

Through its application, sustainable transport has reduced the environmental and social costs of transport in many instances. Notwithstanding or diminishing these achievements are two major limitations. First, these developments on a global scale have not been sufficiently successful in addressing the major environmental failings of urban transport; the gap between theory and practice continues to widen. As Banister (2005), Low (2012), Schiller *et al.* (2010) and many others have argued, at the centre of the problem of environmentally unsustainable mobility is the private motor vehicle and the use of automobility as the solution to modern mobility needs in cities; sustainable transport is not yet the dominant model for urban transport. Second, some activities are contributing little to environmental and social goals (or when broadly assessed have negative impacts) and there is considerable corporate and government 'greenwash' within the field (Romm 2004, Pearse 2012, Auld 2013).

Gössling and Cohen (2014) examined transport 'taboos' in the EU as explanations for the failure of sustainable transport policies. Many, if not all, of their taboos seem applicable to the majority of developed nations. Several are worthy of closer attention in the context of the barriers to community-owned transport. These authors point to basic flaws in public policy, stating that EU transport strategies lack credibility as they are without clearly defined emission targets and effective emissions monitoring; further, these policies emphasise technological solutions that have yet to be developed. They find market-based measures to be socially and economically regressive, in that the greatest effects are on the least wealthy, whilst the incentives for the most wealthy to change their travel behaviour are weak. Such approaches must also work against continuing public subsidies to transport, noting that the costs of the most damaging transport modes are externalised. Furthermore, the authors note that many key issues in transport are ignored, including measures that are proven to reduce environmental impacts. One of the reasons for this outcome is that major stakeholders in the automotive business have successfully lobbied governments to influence policy outcomes to protect their interests. Travel behaviour includes unnecessary trips, they argue, induced by low costs and status-consciousness. However, policies have also failed because they have not come to terms with how high mobility is embedded in 'sociocultural norms' and the effects of changing social structures and relationships that promote greater mobility.

Returning to the earlier critique of sustainable transport, the concerns that the concept is compromised in ways that trade-off environmental protection for other values have been borne out in practice. Sustainable transport is a form of sustainable development as generally understood and practiced; it is premised on protecting and encouraging economic growth. Recognising that the ecological systems that support life on earth have limits to the extent that their constituent natural resources and ecological services can be exploited without being irreparably damaged is a tenet of environmental sustainability. Environmentally sustainable transport, therefore, offers stronger environmental protections than sustainable transport. Protecting long-term environmental and social values from the negative effects of transport systems will require a far more extensive and ecologically grounded approach that is offered under sustainable transport as currently understood.

Part II

Foundations of Community-Owned Transport

4 Collaborative Consumption

Few aspects of urban transport have attracted the recent attention of travellers, businesses and governments, and now transport researchers, as much as the phenomenon of collaborative consumption. Carsharing, ridesharing and bikesharing schemes were almost unheard of until the late 1990s, but have now become a common and routine feature in many people's travelling habits and decisions in cities around the world. This chapter describes this form of mobility provision and its place as a plausible foundation of at least one aspect of community-owned transport.

Sharing in Capitalist Economies

Collaborative consumption is a new term for the old idea and practice of shared (or joint) consumption. This apparent re-discovery is due largely to a new generation of businesses and social institutions based on sharing and no doubt much of this interest is based on the publicity and marketing associated with the commercial aspects of this activity. Without doubt, the advent of the new generation of information and communication technologies has transformed many aspects of contemporary life for governments, businesses and the wider community. Many foresaw the potential of a commercial revolution, and a dot-com boom (and subsequent bust) occurred in the early 2000s. As well as the commercial opportunities, the spread of the new technologies also evoked utopian visions for community and individual empowerment. Now some years down the track, the role and use of personal computing and the Internet have made great impacts in everyday life. However, many of the earlier expectations have been lowered in the wake of experience. New businesses and commercial opportunities have grown and communities and individuals have benefitted, but to some extent the transformative and positive virtues of these changes were oversold.

Within the enthusiasm for this new era, few foresaw one of the key social changes, which at the time was called peer-to-peer commerce, now known widely as collaborative consumption, a term popularised by Botsman and Rogers (2010). This 'sharing economy' now covers a broad spectrum of commercial and non-commercial activities, products and services, and there have arisen a slew of new enterprises and relationships. It covers both collaborative production (such as through design, production and distribution) and consumption, notably through redistribution and shared access. New forms of decentralisation have also emerged in the finance (e.g. crowd sourced financing and peer-to-peer banking) and education sectors (e.g. open education models). *Time* magazine in (March) 2011 called collaborative consumption one of the ten ideas that will change the world. It has been a section of the economy that has attracted high growth in the developed world, with many thousands of companies and millions of individuals now participating.

Sharing is likely as ancient as any other social activity, given that we are a species that lives in groups. Human survival depends on shared access to natural resources and services and industrial society depends further on shared infrastructure, institutions and knowledge. Culturally, the role of sharing is central and it forms part of many social activities and beliefs, such as in religious ceremonies, social ties, family bonding, celebrations and rituals. Not only is sharing a routine part of human ecology and social life; it is also essential in moral systems and codes of conduct and is tied to a range of ethical beliefs. Naming the sharing economy is not merely inscribing a social phenomenon by an academic discipline, although this is partly what the sharing economy appears to denote, but it also marks a change in sharing. That change is a monetisation of sharing that did not previously occur or, if occurring, was not recognised as such.

Collaborative consumption covers a range of activities that Botsman and Rogers (2010) categorise as product service systems, redistribution markets and collaborative lifestyles. Most prominent in media and business interest is the shared use of services and resources under commercial arrangements. Such enterprises generally entail selling access to goods and services on a time basis; i.e. these are essentially rental businesses. What distinguishes these businesses from traditional models are opportunities to derive value from a wide array of products and services without the need for outright purchase, an opportunity characterised as 'access rather than ownership'. Increasingly, this has resulted in an outcome of changes in industrial production, especially the shift to greater technology-based processes, the application of information technology, models of flexible production and supply-chain management and new models of corporate financing. Greater use of such products and services can be found across the economy, in aerospace and defence, automotive, chemical and plastics, construction, food production, healthcare, information technology and communications, industrial products, media, mining and transport sectors.

Then there are businesses dealing with the sale of used goods and non-commercial relationships based on the gift or exchange of used goods (sometimes known as redistribution markets). Perhaps the best-known examples are the on-line auction businesses, eBay and Craigslist. There are also a considerable number of specialist product and service on-line auction business. Additionally, there are on-line exchanges and donation sites, where members exchange unwanted goods and others that offer unwanted goods for free.

Collaborative lifestyles, the third category of collaboration, involves non-commercial exchanges between people and groups who have come together with similar interests, beliefs and values. In contrast to production systems and redistribution systems, both of which can involve corporations and monetised relationships, collaborative lifestyles are within the social realm without economic incentives. Usually based around local communities, there is a growing social movement in such activities as:

- community agriculture;
- community gardens;
- shared housing;
- shared household goods and equipment;
- shared schooling, child care, care-giving and pet care;
- shared tasks and space;
- shared transport;
- shared work and workspace.

Of the forms of collaborative consumption, collaborative lifestyles seem most relevant to our interest in community-owned transport. These collaborative associations involve the donation and exchange of skills and knowledge (skillshare), social network building, time (timebanks) and space (office sharing).

Botsman and Rogers (2010) offer four principles for collaborative consumption: critical mass, idle capacity, belief in the commons and trust between strangers. Critical mass, for the authors, refers to having sufficient participation to maintain viability as size confers sufficient choice of products and services and convenience of use to satisfy consumers. Montreal's bike-share scheme, BIXI, is used to exemplify the concept, describing how 3,000 bicycles and docking stations were not to be more than 1,000 feet (approximately 300 metres) apart in order to optimise utility for users. Critical mass is essential for a second reason, they argue, namely because of the importance of a core group of loyal users who can provide 'social proof' of the worth of the experience. Idle capacity simply refers to unutilised or underutilised assets and services available for use, such as cars sitting idly, empty offices or spare office cubicles, vacant land, specialist or general skills, energy provision and other resources. A belief in the commons is interpreted broadly as people organising themselves to care for common resources (with the authors drawing on the work of Elinor Ostrom), which the authors further endorse for the rewards that cooperation brings to individual participants. Trust is particularly important in collaborative consumption, they argue, because the hierarchal mechanisms of command and control have been removed, 'along with layers of permission, decision-making, and middlemen' and replaced by peer-to-peer platforms and so-called 'transparent' communities.

These commercial services are nearly always based on information provided by Internet access through various personal devices, notably the personal computer and the mobile telephone, the latter of which has achieved high levels of use and ownership worldwide. Markets have opened up globally as a result of these developments, achieving a globalisation of production and consumption at the scale of the individual consumer. Similarly, these technologies have enabled a new form of social networking, connecting people and not-for-profit groups across the world. A major facilitator of these new forms of economic relationships is the electronic payment system, i.e. the phenomenon of e-commerce. Local non-commercial collaborations, however, are not dependent on the Internet and are arranged in a host of other ways.

Collaborative consumption has arisen partly because of the economic advantages that these services can offer over established rival services, notably lower costs and flexibility. In many instances, households and individuals do not face the costs in supplying services facing businesses associated with meeting taxation, financial, regulatory, safety and other responsibilities. Such consumption benefits from the scale of the market that the Internet can access, combined with its ability to allow transactions at costs irrespective of the distances between parties. At the other end of the scale, this consumption also has a strong local emphasis; it is suited to providing goods and services locally at the small scale. This has facilitated the entry of many small providers into particular markets and the challenge for large-scale providers is to find ways to aggregate local demand to achieve the economies of scale they require. Another advantage offered by these arrangements is speed and flexibility; transactions can be arranged and conducted quickly, with electronic payments being a key enabler of business transactions.

Given that a major challenge for businesses and households is the cost of surplus capacity in services and assets, the opportunities to sell such capacity are particularly attractive. It is not surprising that the most popular forms of household-based collaborative capacity involve renting rooms in the family home and selling rides in the household car, given the significance of these items in household budgets and the extent of empty spare rooms and empty seats in the family car, especially when commuting. Co-innovation is another of the features of the shared economy. Rather than products and services being designed and developed within the confines of the corporation, co-innovation involves stakeholders participating in these processes. Consumers, public advocates, interest groups and others contribute to product and service innovation.

Perhaps one of the reasons why this phenomenon was largely unpredicted by the technology boosters was that it represents more than a technological change, but reflects several social trends. In the contemporary era of austerity across much of the developed world (such as marked by the lingering effects of the 2007–08 Global Financial Crisis and the on-going European sovereign debt crisis), the lower and middle classes have seen effectively stable real incomes for many years, household indebtedness increase, the rise in part-time and causal employment, and household savings levels decline. At the domestic scale, the sum of these financial conditions has created incentives to increase income and to lower the costs of consumption. Selling or trading surplus household goods, renting spare rooms for short-term stays and looking for casual employment for cash payments are household financial strategies once used as austerity measures; now they are called collaborative consumption and are facilitated by the Internet, but may be motivated by the same fiscal causes.

There have been many implications for traditional businesses in these new enterprises and market changes and there is a burgeoning business literature on the sharing economy, with some describing it as the future of global economics (Gansky 2010). Most of the larger firms in collaborative consumption are new entrants (typically backed by venture capitalists). In some senses, several industry sectors are marked by competition between traditional businesses and collaborative firms, such as the hospitality sector (hotels vs. room rental), publishing and, significantly from the perspective of this volume, transport (public transport and taxis vs. ridesharing). There is also a countering effect, whereby the expansion of collaborative consumption is identified with the expansion of social activity that is kept out of the reach of markets, as evidenced with the emergence of the Transition Towns, de-growth economics, cooperative economies, the solidarity economy and related movements (Kostakis and Bauwens 2014). Such developments are not following capitalist models of production and consumption, but are built on principles of equity and cooperation, sometimes labelled as commons-based peer production.

Advocates of collaborative economy have made much of the role of trust in these relationships, to the extent that claims have been made that the digital economy has enabled greater trust within the broader community and wider world. However, such claims are difficult to substantiate with evidence. Further, it is claimed that the sharing economy has other socially desirable attributes, notably that it promotes a decentralisation of power and wealth. This may be true, but such claims are also very difficult to substantiate. Any flattening of the social distribution of wealth must be offset by the formation and growth of major corporations providing shared economy services: Airbnb, for example, is the major home rental service and is capitalised now in the tens of billions of US dollars and is still growing. Inequality in income distribution has generally been increasing in OECD nations in recent decades, so if collaborative consumption were equalizing incomes, it would be running against prevailing trends.

Collaborative consumption is a disruptive innovation; it can transform markets, change paradigms of production and consumption, and cause institutional change. Although it is often depicted as primarily empowering alternative producers to enter the marketplace at the margins and compete with established mainstream corporations, there are consequences for those with interests in established businesses; while renting out a household room might only slightly diminish the profits of an established local hotel, it could also cost the job of a hotel employee. Collaborative consumption entails other social, economic and environmental changes, not all of which are positive, as discussed below.

Environmental Implications

Collaborative consumption may have a role to play in environmental protection by promoting eco-innovation, reducing natural resource consumption and lowering pollution related to consumption. 'Repair, reuse and recycle', an earlier mantra of green consumption, can be fitted into systems of collaborative consumption.

Such ideas fit with the eco-efficiency and natural capitalism movements (e.g. von Weizsächer *et al.* 1998, McDonough and Braungart 2009). On the surface, collaborative consumption runs counter to the tenets of mass consumption, one of the bedrocks of industrial capitalist societies. Potentially, the sharing of consumptive goods and services reduces the need for individual purchases as one purchase can serve the needs of multiple sharing consumers. Potential disruption, however, goes deeper than reducing sales as collaborative consumption challenges the validity of planned obsolescence as a foundation strategy of consumer goods and services producers. Product design has long featured attributes to foster obsolescence through strategies that limit product life, make replacing impaired products more attractive than repair, by promoting fashionability and ensuring individual component developments render existing assemblages redundant. These features are challenged by eco-efficiency designs and practices.

Increasing globalisation of production has seen the chains of industrial production lengthened and fragmented in an on-going drive to find locations of lower production costs and, less well documented, locations for the disposal of industrial wastes. Globalisation and rising prosperity in the newly industrialised and developing nations is driving global consumption upwards. Reckonings of the environmental costs of these trends are partly behind the global movement in sustainable development and global environmental awareness. Estimations of the ecological costs of mass consumption, such as by the ecological footprint measurements, demonstrate the biological and biophysical limits to economic growth. One response has been the 'green consumption' movement that, in essence, promotes consumption with an awareness of the environmental and social implications of product choices, uses and disposal. Ideas for reforming economic markets towards environmental goals, such as those of natural capitalism, have green consumption as a central feature (von Weizsächer *et al.* 1998). It follows that collaborative consumption has been promoted as strategy to promote such green consumption (Botsman and Rogers 2010). Collaborative consumption is a relative newcomer to this movement, but does have some prima face credibility as a response for curbing mass consumption.

As to its environmental benefits and its association with the cradle-to-cradle notion of closing the loop of production and consumption, while there are some outstanding case studies, there are still relatively few thorough and wider-ranging evaluations of the environmental costs and benefits of collaborative consumption. Leismann *et al.* (2013) studied the environmental impacts of collaborative consumption and found on the positive side of use-rather-than-own schemes, an extension of product life and use of durable products, use of energy efficient appliances, maximisation of device use, consideration of technological/ecological progress, promotion of recyclable construction design and economies of scale and benefits of specialization. Negative aspects of collaborative consumption were found to include greater wear and tear than normal use, overuse, accelerated withdrawal of rental products that are still in working order, too long use of inefficient appliances, additional resource consumption to extend useful life and additional transportation (Leismann *et al.* 2013). Clearly, the environmental consequences of collaborative consumption are a complex matter where simplistic claims of environmental benefits cannot be taken at face value. Establishing the environmental credentials of collaborative consumption remains a nascent field and basic questions still remain open as to its environmental effects.

A Dark Side to Collaborative Consumption?

While there is much support and enthusiasm for this new realm of economic and social activity – if not outright boosterism – it has a dark side and there are serious grounds for concern over collaborative consumption, especially regarding product service systems and the possibilities of increased material consumption. Collaborative consumption can have the aforementioned environmental benefits associated with reducing mass consumption and increasing the efficiency of material and energy use, but these may be mostly seen as side benefits to motivations that are primarily economic. Declining real household income for many in OECD nations (in the wake of long-term increases in household income inequality in the OECD nations over several decades) is likely to be a significant influence promoting widespread involvement in the sharing economy (Summers and Balls 2015). Against the lauding of the sharing economy for reducing the influence of big business through boosting market competition is the loss of employment from big business, so that the peer-to-peer economy may be damaging the security of the paid workforce. Potentially, the sharing economy also diminishes the corporate tax base, with broader implications for funding of public sector activities. A major concern is the creation of a new form of low-paid employment where people are employed at the lowest wages for the smallest jobs and all without the set of protections provided by 'proper' jobs, such as health and safety requirements, accident insurance, health insurance and superannuation (Summers and Balls 2015). There is no minimum wage for such informally arranged employment; rather than being peer-to-peer, these businesses may be no more than a broadening of what used to be known as the underground economy marked by exploitation of the least advantaged. In some respects, the sharing economy is based on shifting the risks of production and consumption onto those with the fewest resources and capacities for understanding and avoiding the risks associated with economic activity. Rather than ushering in the future of paid work, collaborative consumption may be facilitating the return of the worst aspects of early and unregulated industrial capitalism.

A critical point here is whether or not collaborative consumption is being used as a strategy to lower the overall costs of labour or whether it is one to increase community access to underutilised resources. Looking broadly across collaborative consumption as a whole, it is clear that a large number of schemes are, in all but name, a new form of the informal sector economy within developed nations. Although the rhetoric of the new productive service business evokes the image of the small and dynamic firm outperforming the large traditional corporation, much of the activity is decidedly less heroic and involves workers with few other prospects trying to low-bid each other into piecework cash employment.

A major factor in the cost advantages of these schemes is not efficiency per se, but the avoidance of the normal overheads of legitimate businesses (such as taxes on earnings, superannuation contributions, health insurance and compliance with health and safety regulations). It seems highly likely that a large swath of the 'task rabbit' schemes, having come under closer scrutiny of the state, will be declared illegal or will involve the participants being brought into the more formal and regulated labour market. In the absence of regulatory extension to cover these new employment contracts, which are noticeably without formal agreements, it will be left to the courts to determine responsibilities in cases of dispute. Online reporting and mutual surveillance through such virtual means are promoted as a kind of community self-regulation, but it seems naive to argue that monitoring of the participants' reputations is sufficient to prevent malfeasance and is an effective substitute for health and safety regulation, employment arbitration, workplace safety assessment and the like.

Many jurisdictions oppose collaborative consumption firms and operators operating product services businesses. While supporters of the firms point to the role of disruptive industries as sources of change, economic growth and innovation and suggest that opposition stems from vested interests, governments and public administrators have found some of the new emerging collaborative consumption firms to be law breakers. A report by the New York state Attorney General, for example, found that 72 per cent of short-term rentals violated laws against short-term stays (Schneiderman 2014). Some cities moved to enforce existing laws governing the prominent sharing industries in hospitality and transport, other have moved to enact new laws while other have sought ways to incorporate these new firms into the legal and regulatory framework. London scrapped laws against short-term visitor stays and San Francisco allowed short-term rentals but sought hotel taxes from the new operators, exemplifying strategies of appeasement, while a number of cities have banned collaborative consumption taxi services (such as Uber), as in Germany and Canada.

Community-based activity that is largely without commercial interests, however, seems a considerably more benign realm of activity than product service businesses. There are a number of community-based schemes that do not impinge on anyone's livelihood or conditions of work. In this sense, community ownership is not strictly a disruptive business model that seeks to subvert established commercial institutions and firms, but rather is a practice that builds new forms of enterprises without necessarily undermining existing firms, such as public transport providers.

Transport as Collaborative Consumption

Transport has been prominent in collaborative consumption. In a coincidence, one of the first articles (perhaps the first) on collaborative consumption, by Felson and Spaeth (1978) in the *American Behavioural Scientist*, dealt with carsharing, although the term has only come of age in recent years. Prettenhaler and Steininger (1999) considered the potential of carsharing at the point at which the concept was beginning to take off. Sharing transport offers many of the benefits of private vehicle use but without some of the costs, such as those of depreciation, operating, insuring, licencing, maintaining and parking your own vehicle. There are several forms of collaborative consumption in contemporary transport, with the possibility that more may emerge over time. That which is closest to private vehicle ownership is carsharing, which is essentially short-term vehicle leasing. Ridesharing is the other major form of collaborative transport, which is either a form of a taxi service provided by a private operator or the use of a communal vehicle (usually on journey to a common destination for the passengers). There has been spectacular growth in these industries in the last few years; some cities report that the total services now rival those of conventional taxi services. Uber, the ridesharing company, is less than a decade old and now operates in 250 cities worldwide.

Despite the relative newness of widespread carsharing businesses and not-for-profits and uncertainty over their eventual contribution to urban mobility as whole, the concept has been widely promoted as one with a role to play in sustainable transport, notably in the last few years (Banister 2005, Goldman and Gorman 2006, Urry 2008, WBCSD 2004a, Schiller *et al.* 2010). For some transport planners, carsharing is an option to be encouraged through specific support and incorporation into transport plans (Tumlin 2011). More specifically, Jussiant (2002) envisions the environmental benefits of carsharing will be achieved through integrating carsharing with other transport modes. More broadly, urban planning reformers and eco-city advocates see carsharing, ridesharing and bike-sharing as contributing to efforts for reducing the environmental costs of urban activity and improving urban amenity (Beatley 1999, Roseland 2012). Critics of the car and car-based transport systems have also seen carsharing as one route to weakening car ownership, level of usage and car-dependency (Dennis and Urry 2009) and having a role in weakening the 'culture of cars' (Wright and Curtis 2005). Few have gone as far as the architect Moshe Safdie (1998) however, who, in *The City After the Automobile*, called for replacing private car ownership with mass carsharing of electric vehicles. As part of lifestyle strategy to live more economically and sustainably in community settings, a wide array of works promote collaborative consumption of transport as a worthy practice (e.g. Chiras and Wann 2003).

Carsharing

Carsharing or car-pooling (aka 'car clubs' in the UK) has become widespread, as either a commercial venture, corporate carpool, public body, cooperative or on a casual contract basis. Several conditions distinguish carsharing from other forms of car use: it is a membership-based activity where members (as individuals or firms) have access to vehicles from a fleet for use on an (usually) hourly basis and pay a monthly or annual membership fee. Before the year 2000, there were few such schemes anywhere (although the contemporary concept was floated in 1969 by Fishman and Wabe [1969]). An earlier US term for carsharing was 'community garages', which perhaps better captures the concept of the common use of a vehicle pool.

There is also what might be termed 'informal car sharing' in which there are no third parties involved in vehicle sharing (i.e. a firm, legal entity or formal institution), so that formal and informal car sharing differ significantly. Informal carsharing is more likely to be used by close-knit communities, locations with low car ownership, poor public transport and few local facilities, places where there are popular destinations and trip patterns suited to journey sharing and where there are incentives for sharing vehicles (Bonsall *et al.* 2002). Organised car sharing is promoted by the same conditions as informal carsharing, but is enabled by such factors as a larger population of newcomers, new travel patterns, the loss of public transport services or where employers encourage car sharing (Bonsall *et al.* 2002).

Operations are now spread around the world, with North America being the major market. By 2012, carsharing was operating in 26 nations and in some 1,100 cities, with over 1.7 million members and a combined fleet of over 43,000 vehicles; North America has around 50 per cent of the members and Europe, around 40 per cent, although Europe has a greater number of vehicle fleets (Shaheen and Cohen 2013). There were many, typically relatively short-lived, precursors to contemporary carsharing in Europe and the US (Shaheen *et al.* 1999, Millard-Ball *et al.* 2005). Some of the better-known and longer-lasting schemes were a Zurich housing cooperative in the 1940s that included car sharing, the Sefage programme (selbstfahregemeinschaft); Montpelier had the ProcoTip programme in the early 1970s and Britain had its Green Cars programme between 1977–84. Other early efforts were Amsterdam's witkar, using electric vehicles (1974–88) and several schemes in Sweden, including bilpoolen in Lund (1976–79), Vivallabil in Örebro (1983–98) and in Gothenburg (1985–90). ATG (Auto Teilet Genossenschaft) and ShareCom, two Swiss carshare cooperatives, merged to form Mobility Carsharing Switzerland in 1987. While ATG was simply a carsharing cooperative, ShareCom covered the sharing of houses, tools and other items, while StaAuto was started in Germany in 1988 as a university research project. In the US, Mobility Enterprises (1983–86) and STAR (Short-Term Auto Rental (1983–85) were begun as experiments. In 1998, Carsharing Portland begun and soon after, Flexcar and Zipcar began. CarSharing Portland was bought by Flexcar in 2001 and Flexcar and Zipcar merged in 2007.

In essence, formal carsharing is a form of car rental on a short-term, but it differs from conventional car rental in a couple of key ways. Carsharing entails paid or unpaid membership of a corporate service provider or other institution; joining usually involves providing proof of driving qualifications and opening an account, after which pre-approval to drive is given. Self-service is a feature of carsharing, with customers making their own on-line bookings, car collection and car return. Vehicle collection and return is to exclusive and designated parking locations. In most cases, the vehicles are available across a variety of locations, with many schemes locating vehicles in close proximity to public transport. Usage is charged on a time basis, typically hourly and there is no charge for fuel use.

Vehicles are also not typically fuelled or cleaned between users. Conventional car rental involves separate booking procedures for every hire, hiring is usually on a daily basis, charging schemes usually include a number of variable and separate costs (such as period of hire, mileage and insurance) and there are limited locations for vehicle returns.

A range of studies into carsharing in different locations has revealed many common features across different firms and services. Users tend to have middle to high incomes, be aged in their 30s and 40s, with higher levels of formal education (Millard-Ball *et al.* 2005). Uptake of carsharing is concentrated in city centres and urban locations where there is relatively little car dependency. Given that carsharing has been taken up most rapidly in North America and Europe and amongst those who are better off, it is a mobility choice for those that will often have a range of mobility choices, including car ownership. Members of carshare schemes have lower vehicle ownership levels. Findings from the Transit Cooperative Research Programme also found members to have a high regard for social and environmental issues (Millard-Ball *et al.* 2005). Motivation for joining carsharing schemes is usually simply the occasional need for access to motor vehicles for relatively short and perhaps one-way trips (for which conventional vehicle rental is inconvenient and more expensive) and available services nearby, as the use of services tends to be higher by those living closer to car stations (Katzev 2003). Usage of carshare vehicles centres on personal trips for shopping (especially for groceries), personal business and recreation; notably, there is very little use for commuting (Millard-Ball *et al.* 2005). Household users made up nearly all the initial markets since the 1990s, but in recent years, businesses have begun to use carshare services.

There has been considerable interest by researchers and policy-makers in carsharing, perhaps because of its novelty value (as it accounts for only relatively few urban trips), but almost certainly because it adds a new form of urban mobility, albeit using the ubiquitous car. Members of carsharing schemes use carshare vehicles on an occasional basis at low rates, so that it supplements other private, public and active transport. Many carshare members access share cars using public transport. Carshare can serve as the 'missing link' for city dwellers who have only occasional need for using a car and find carsharing attractive for practical, economic and environmental reasons.

Whether or not carsharing reduces individual or household environmental impacts is a key question. Providing car mobility through carsharing increases car use and its associated environmental impacts, but it can also reduce these impacts as a result of several effects on car ownership and use, producing net environmental benefits. Carshare users in the US have overall one-half the rate of car ownership than the wider community and each sharecar vehicle takes the place of around 15 vehicles on the road (Millard-Ball *et al.* 2005). Carsharing enables households to divest themselves of a second or third motor vehicle (Martin and Shaheen 2010a). Millard-Ball *et al.* (2005) found that, on average, 20 per cent of carshare members had forgone new car purchase. Carshare use varies greatly by user, but the rates of carshare use are typically low in comparison to the rate of car owners; two trips a month was found to be typical in one review and carshare drivers cover 40 per cent less distance after joining a carshare scheme (Millard-Ball *et al.* 2005). By reducing private car use, carsharing shifts a portion of mobility to public and active transport. In sum, carsharing schemes were found to reduce observed North American greenhouse gas emissions by 0.6 tonnes of greenhouse gases per person per annum due to reduced driving and 0.8 tonnes if the full effect of carshare use is assessed (Martin and Shaheen 2010b). An earlier European study found that carsharing participants reduced their transport carbon dioxide emissions by 40–50 per cent (Ryden and Morin 2005).

Certainly, carsharing is attractive economically compared to private car ownership and compared with conventional hire cars that are expensive and impractical for short-term use. Vehicle choice by carshare operators is typically of smaller, fuel efficient vehicles and hybrid vehicles, offering some emissions savings over the average of the urban fleet of private cars. Millard-Ball *et al.* (2005) found strong evidence to support claims that carsharing produces cost savings to households, increases mobility and convenience, reduces parking demands and uses more fuel-efficient vehicles. Success in establishing carsharing enterprises often depends on the assistance of partner organisations, whether as local government or state transport operators, land developers, businesses and tertiary education centres. There are many forms to this assistance, such as local governments providing dedicated street parking, including carsharing in transport planning and marketing programmes or direct assistance from local businesses in facilitating and promoting carshare use (Enoch 2002). Despite the rapid growth in carsharing and its prominence in discussions about emerging industries, it is worth remembering that it still only involves a few tens of thousands of vehicles in a global fleet that exceeds a billion vehicles.

Ridesharing

Perhaps the most newsworthy development in worldwide urban transport has been the rise of new commercial ridesharing businesses. Essentially, it is a private, on-demand taxi service provided by private and independent drivers operated by the rideshare company. This new industry is essentially a revival of the informal transport sector, with private car owners and drivers providing taxi services to individuals through a private company that manages the accounts and operations using mobile telephone and Internet software. Trips are pre-arranged, are usually single and non-recurring; drivers and passengers are matched using automatic matching systems operated by the journey share company. This activity goes by several names: ridesharing, lift sharing and journey sharing. Basically, the commercial (and voluntary) operations follow a basic common model. A service provider operates an on-line service through which car drivers and passengers exchange information necessary for booking a ride and riders pay the company using their electronic account, from which the drivers are paid, less the commission charged by the provider. Charges can be based on distance or time. Critically, the service provider only facilitates the transactions, the car and driver are independent operators registered with the company. A percentage of the fare is taken by the company and this provides its primary revenue source. Critically, the costs and risks of providing the service are borne by the driver. Passengers provide ratings of their experiences that are made available to all users, so that drivers acquire an online profile of their service.

Many of the major corporations in the market have already become household names in their local markets, such as Uber (number 1 in the US market) and Lyft based in the US, Kuadi Dache (number 1 in China) and Didi Dache in China, GrabTaxi across Southeast Asia (based in Malaysia) and Ola and TaxiforSure in India. Other popular journey sharing enterprises, mostly based in the US, are PuckupPal, GoLoco, Zimride, Ridester and Craigslist. Growth has been spectacular, notably in the case of Uber, the best-known of these firms in the developed world. Uber was founded in 2009 in San Francisco and was valued at US$40 billion in late 2014, a doubling of its value from 6 months earlier (Johnson 2014); it claims to be operating in 50 nations. Liftshare, the UK's largest carsharing network, is an interesting example of such firms as it is a non-for-profit service that bases its business model on developing car share schemes for private clients.

Sharing cars has always occurred informally, but with the rise of the Internet and widespread mobile phone use, specialised software to facilitate ridesharing has enabled a new generation of activity, both on a voluntary basis and more noticeably, as a new business economy. What has attracted so much attention to these new businesses is their very high rates of growth in usage, international spread and, in the case of some firms, very high market valuations. Also attracting attention has been the considerable controversy over the industry, with many such services illegally competing with established taxi companies with exclusive operating rights, as discussed below.

These services have several benefits over existing taxi and shuttle services, many of which stem from providing solutions to many long-standing complaints from regular taxi users in many cities. Commercial ridesharing provides on-demand services and many users find this, or perceive it to be so, more reliable than using taxi telephone booking services or hailing cars in the street. Rideshare journeys are typically cheaper than cabs and the drivers do not have to be tipped. Tipping of conventional taxi drivers is often preferred in cash, which can be problematic with so many daily purchases now done electronically. Ridesharing fits with the model of cashless commercial transactions that is also becoming increasingly widespread and popular. Fare advice can be obtained prior to the journey and this alleviates concerns that taxis may inflate fares or take longer routes. Most commercial ridesharing companies handle all transactions on-line, including booking, payment, information about the driver and vehicle identity.

According to the ridesharing firms, commercial journey share trips are cheaper than regular taxis and the drivers are better paid than taxi drivers. Assessing such claims requires consideration of other perspectives. Since the basis of ridesharing companies is independent drivers using their own vehicles, such operators would need to be more rewarded than a driver working for a taxi company in order to cover their vehicle and associated costs. Whether or not there is a financial benefit to both drivers and passengers is difficult to determine and certainly there are anecdotal accounts questioning the claims of Uber and other companies regarding fares and driver incomes. As to the scale of the new industry, this is also difficult to determine at this time for there is little independent data available from these privately owned firms. Writing for *Forbes* magazine, a San Francisco-based reporter suggested that Uber journey share vehicles could exceed that city's regular taxi fleet in two years; in 2014, he reported that there were over 1,000 journey sharing vehicles on the streets at peak hour and 1,600 regular taxis (Rogowsky 2014).

Cairns (2011) in a review of different car ownership models in the UK, identified two approaches to encouraging ridesharing: 1) Direct encouragement through national or international schemes, local and regional schemes, and employer and school schemes and 2) Preferential conditions for shared journeys, such as high-occupancy vehicle lanes and preferential parking for ridesharing vehicles. City programmes to promote higher car occupancy have been around since the 1970s and have often been unsuccessful or disappointing, with two exceptions: those of company-based programmes and high-occupancy vehicle lanes (although 'high' can mean the driver and one passenger in some places), both of which have produced successful outcomes given the right conditions. With the arrival of high-volume organised ridesharing, however, it is clear that at least some of the existing presumptions about ridesharing will need to be revised.

As several studies have made clear, ridesharing reduces the environmental costs of car transport if a shared journey is assumed to be replacing two separate motor vehicle journeys. In many developed world cities, informal ridesharing has declined over recent decades, prompting Levosky and Greenburg (2001), amongst others, to consider the environmental benefits of promoting organised ridesharing. In 2005, the International Energy Agency (IEA 2005) showed considerable reductions in transport fuel use (ranging from regional savings of 17–30 per cent) by adding an additional person to every urban car trip. Studies of ridesharing have found savings in transport emissions (Caulfield 2009, Jacobson and King 2009). A full reckoning of the environmental effects of ridesharing will have to take into account the rapid evolution of these businesses. Oddly, the commercial rideshare companies tend not to identify themselves as transport businesses, per se, but as companies that facilitate and arrange ridesharing using on-line technologies, i.e. these firms self-identify as service providers. It is the thousands of individual drivers, typically using their own vehicles, offering their services to the public who are partnered with the rideshare companies that are providing the mobility service.

Bikesharing

Known variously as bicycle sharing schemes, public use bicycles and smart bikes, these city-based services provide short-term rental bicycles from a network of bicycle stations. Using specially designed and distinctly painted bicycles, these systems operate on smart cards or credit cards for temporary or longer period membership, with a use pricing structure designed for short trips (often with the first 30 minutes free). Bicycles can be taken from a docking station and returned to any other station location, with the system oriented towards a high volume of short trips and maintaining a large fleet of available bicycles. Bikesharing schemes have become fashionable amongst the world's prominent cities and although sometimes derided by transport planning professionals as a recreational novelty, there is now evidence that they are providing a genuine mobility service. Arguably, bikesharing schemes are the fastest-growing transport systems worldwide. Although there have been a relatively small number of such schemes in the past, they tended not to last very long; the current generation of bicycle sharing schemes are numerous, occur across the world and in a few cases, are very large. In 2014, there were some 712 cities with bikesharing schemes across five continents with a combined fleet exceeding 800,000 bicycles (Shaheen *et al.* 2014). To this can be added an unknown number of corporate and university bikesharing schemes.

DeMaio (2009) and Shaheen *et al.* (2010) identify three historical phases of bikesharing schemes, beginning with the free public use schemes, exemplified by Amsterdam's 'white bicycle' scheme in 1965; abuse and theft caused this public scheme to collapse within days. This was followed by modest bikesharing schemes in a few Danish towns in the early 1990s and, in 1995, a larger scheme in Copenhagen. Marking this second generation, the coin-deposit phase, featuring use of coin deposits, utilitarian bicycles and fixed collection/return docking stations. Bicycle theft remained a problem and the solution of customer tracking gave rise to a third and contemporary model of bicycle sharing, that which is based on information technologies. Technological innovations in this next phase included electronically locking racks, smart card access, centralised management systems and mobile phone access. Several schemes emerged utilising these improvements, notably Rennes Vélo á la Carte in France in 1998 with features that have now become standard: a network of stations, automatic docking, real-time information systems and cashless and smart card technologies. However, it was the success of Paris's Velib system and Barcelona's Biking schemes that brought the concept to wider attention (Midgley 2011), partly because of their scale (e.g. Paris began with 7000 bicycles). Although this recent generation of bikesharing schemes was European in origin, it is now a worldwide phenomenon, with some spectacular schemes in China, with a very high rate of growth in Europe after the mid-2000s and slightly later in Asia-Pacific. Wuhan and Hangzhou in China have the world's largest scheme, with some 90,000 bicycles in Wuham; Hangzhou has 60,600 bicycles and 2,416 fixed stations, achieving up to 320,000 uses on a single day (Shaheen *et al.* 2011).

Many factors have been identified behind the successful growth of the bikesharing schemes, not least that the schemes have been able to address many earlier problems and have been the beneficiaries of the information and communication technology revolution. Users have been attracted to these schemes because of their 24/7 availability, the speed of short city trips by bicycle, relatively low-cost subscriptions and their environmental and financial attractiveness. Successful operations have much to do with achieving a network of docking stations in locations of high-potential demand and achieving a sufficient density of stations (such as having only 300 metres between stations). One planning guide recommends a minimum system coverage of 10 square kilometres, with 10–16 stations per square kilometre and 10–30 bicycles for every 1,000 residents covered by the scheme (ITDP 2013). Other features of successful schemes include comfortable, commuter style bicycles, automated locking systems, wireless track systems of bicycles, real-time monitoring of docking stations, real-time user information available through the Internet and pricing that encourages short trips and maximizes the daily trip use per bicycle (ITDP 2013).

There is a considerable range of business models for bikesharing schemes, from the corporate, public to the non-profit models (DeMaio 2009, Shaheen *et al.* 2010, ITDP 2013). Corporate models involve advertising companies that provide and operate the scheme for advertising rights on bikesharing stations, street furniture and billboards from which they generate revenue in addition to membership fees (e.g. SmartBike in the US and Cyclocity in France). For-profit corporate ownership is a more conventional corporate ownership model based on revenue from membership (and usually advertising rights), (e.g. nextbike in Germany). Local or city government authorities own and operate some schemes (or contract this task to others), providing municipal funding, membership fees and advertising revenue as revenue sources (e.g. OV-fiets in the Netherlands and Nubija in South Korea). Public transport operators have business models based on government subsidy as well as income from the usual membership fees and advertising sources; oftentimes, the bikesharing scheme is formally incorporated into the public transport system, such as through a common ticketing and payment system (e.g. Hangzhou Public Bicycle in China and Call a Bike in Germany). Finally, there are the non-profit schemes that are usually supported by local governments and public agencies and employ a variety of business models and income sources, including public-private partnerships, bank loans, and local funding to supplement membership fee income (e.g. BIXI in Montreal, Call a Bike in Germany), corporations and not-for-profit groups (e.g. Bycyklen in Canada, Wuhan Public Bicycle in China and Hourbike in the UK).

Shaheen *et al.* (2010) identify the benefits of bikesharing as: 1) Providing additional mobility options, 2) Cost savings, 3) Lower costs for accessing other modes, 4) Reducing road congestion, 5) Increasing the use of public transport and alternative modes, 6) Health benefits and 7) Increased environmental awareness. These authors also note the role of bikesharing in addressing the 'last mile' problem (i.e. the final short distance between origin and final destination), noting that: 'Thus, bikesharing has the potential to play an important role in bridging the gap in existing transportation networks, as well as encouraging individuals to use multiple transportation modes' (2010, p. 159). Other benefits include those resulting from reducing car use and increased public transport, so that pollution and other social and environmental costs of road transport are lower, overall cycling rates can be lifted and there can be local investment opportunities associated with bicycle sharing schemes. Cities are also taking up bicycle sharing as a way to enhance their marketing and image; in 2007, for example, the Paris Vélib system won the 2007 British Guild of Travel Writers Best Worldwide Tourism project.

As for the impacts of bikesharing schemes on transport sustainability, the success to date is modest according to a small number of studies, with bicycle use largely displacing public transport use (Midgely 2011). Schemes with high rates of use cover substantial distances as total vehicle kilometres travelled, which if covered by private cars would produce high volumes of emissions, but such savings only accrue if bikeshare use entirely substitutes for car use, which is not the case. A study of Montreal's system found that only 2 per cent of its users would have used a car instead of cycling; in Barcelona the shift was 10 per cent (Midgely 2011). A review of a wider range of user surveys in North America found higher reported reductions in car use by bikesharing scheme users, with higher reported levels of walking (Shaheen *et al.* 2014). Another review by Fishman *et al.* (2013) of bikesharing research found that the majority of scheme users are using bicycles as substitutes for the more sustainable transport modes, rather than for the private motorcar. Significantly, although bikesharing is being substituted for public transport use, there is evidence in many cities that the integration of bikesharing and public transport promotes the use of both modes (Fishman *et al.* 2013). Bikesharing, however, clearly does have the potential to reduce vehicular emissions when used as an alternative to private car use. A study by Goodman *et al.* (2014) found evidence that a bikesharing scheme in London could promote greater cycling by playing a role in 'normalising' cycling.

Success, however, has brought problems to some cities, with the opportunities for further growth limited by available street space for additional docking stations, high use rates diminishing the availability of bicycles and the logistics of balancing the docking spaces and bicycle availability. Ultimately, the influence of bicycle sharing in reducing car use will be fully revealed in longer time frames. Even in the few years of operation to date, these contemporary schemes appear to be influential in promoting further investment in cycling infrastructure (bicycle lanes, bicycle paths and bicycle parking) and in lifting the overall level of cycling by attracting previous non-cyclists (ITDP 2013). Given that high rates of cycling are often associated with higher levels of cycling infrastructure, notably across northern Europe, there may well be a virtuous circle between popular bikesharing schemes, city investment in cycling infrastructure and higher cycling rates.

Integration between Collaborative Consumption and Traditional Transport Providers

Car manufacturers have taken up the challenges of collaborative consumption responding with several different strategies, including integrating their businesses into the collaborative consumption economy. This integration takes several forms, notably designing vehicles that can be shared between multiple users and associated vehicle fleet management systems. A relevant example is Toyota's 3-year trial (which started in October 2014) in conjunction with the Grenoble city authorities who operate the public transport services, the French energy company

EDF and a local carshare company, using Toyota 'ultra-compact' electric vehicle, the i-ROAD (which is a two-seat three-wheeler). This is a carsharing scheme with 70 vehicles and around 30 docking stations, sources for charging, that are co-located with public transport stops. This scheme will be integrated with an existing carsharing operator, Citélib.

Essentially, the logic of this experimental system is relatively simple; public transport users will be able to hire an electric vehicle to complete the remaining short leg of their journey. Such small electric vehicles can have less environmental impact than conventional cars and assist in reducing road congestion (the manufacturer's claim that each vehicle occupies one-quarter of the space of a conventional car since the i-ROAD is only 850 mm wide and a little over 2.4 metres long). This system draws on collaborative consumption in several ways; using smart phone technology it offers a 'multi-modal navigation system' that provides real time information for users on the availability of vehicles, traffic conditions and public transport serves and makes route recommendations with a mix of public transport and electric vehicle options, so that users can undertake their own journey planning based on current conditions. Similar to existing carsharing schemes, this system uses a smart phone application; for vehicle booking the hiring transaction is carried out automatically. It is the company's second such 'Ha: Mo' project, the first being in Japan (the moniker stands for 'harmonious transport').

A second strategy of the established car industries involves developing product services that imitate aspects of the product service systems of collaborative consumption. This is exemplified by Peugeot's programme European MU mobility service whereby the car manufacturer offers a rental service that covers cars, electric vehicles, vans, scooters (including an electric motor model called the e-VIVACITY), bicycles and accessories. Peugeot's model combines the familiar car rental business with elements of carsharing businesses, with the option of a much wider range of vehicles. An obvious downside is that the system is run from the manufacturer's retail outlets, so that what it makes up for in flexibility of mode choice it certainly loses in the limited options for collection and return; furthermore, the rental periods are not hourly, but usually a minimum of half a day. It may just be an extended marketing exercise at this stage, rather than offering something innovative in transport. Other car manufacturers have begun carsharing businesses, such as Ford's Ford2go operating in Europe.

There has also been a much simpler and traditional corporate response to collaborative consumption, namely being rival businesses. Car manufacturers have been forming their own carshare businesses that exclusively use their own vehicles. In recent years, in Germany for instance, Daimler AG formed car2go, BMW and Sixt entered a joint venture to form DriveNow and VW formed Quicar.

Idle Vehicle Capacity

Within every city there is a wide range of usage of the vehicles within the vehicle fleet. Commercial vehicles may be in almost constant motion, such as taxis and delivery vehicles, and have high usage per annum. At the other end of the scale are a great number of vehicles that are only used for a relatively short period in every 24-hour cycle. In OECD nations, the typical vehicle loading is less than two passengers, meaning that for most cars and for most trips, the only occupant is the vehicle driver.

From an environmental perspective, the less a vehicle is used, the less pollution it creates (more or less) and conversely, the greater the number of passengers, the less the per capita environmental costs per trip. From another perspective, when there are significant numbers of people without access to transport and a large number of idle motor vehicles and empty seats in vehicles in use, there is asymmetry between needs and resources. Income differences in capitalist economies explain much of this phenomenon; those without access to mobility are usually on lower incomes or are welfare recipients. Wealthier individuals and households can simply purchase mobility (and usually as a relatively small proportion of their net worth). Nationally, this effect is reflected in national income; cities in wealthier nations are dominated by high levels of car ownership and mobility is largely achieved through motor vehicle ownership (with the Western and Northern European cities with high public transport use being the exception, although car ownership is often still quite high). Differences in mobility, therefore, can be explained to a significant degree by differences in socioeconomic status.

Collective use of motor vehicle capacity offers a way to address these effects and does not require changing social wealth distribution. Collaborative consumption can provide access to idle and under-filled motor vehicles and thereby obviate the need for vehicle ownership; this re-casts the problem of mobility in car-dependent societies by breaking the nexus between mobility and vehicle ownership. Access to such mobility brings benefits for all socioeconomic groups, but importantly improves mobility for the transport-disadvantaged. Citywide, sharing of the motor vehicle fleet reduces the need for motor vehicle ownership, lowering total fleet size and for households that reduce their motor vehicle ownership, frees capital that was formally tied up with ownership.

Increasing passenger travels by car will increase transport pollution in the absence of any countervailing actions. A number of options are available for this purpose, as described in Chapter 3, including greater ridesharing (i.e. more passengers per car) that can offset increases in motor vehicle use. Reducing the total vehicle fleet reduces its ecological footprint because the embodied emissions in each vehicle are avoided. Incentives through public policy can also be used to promote the use of more energy efficient vehicles and other vehicle-based approaches.

Reactions against Carsharing

Commercial carsharing, ridesharing and bikesharing schemes are disruptive innovations that displace established business models and firms (notably the car-hiring and taxi businesses). Success by these new firms is partly due to advantages over the existing firms and their services, but critically also by creating new markets and services that the existing firms were not meeting. As described above, one response by the established vehicle hiring businesses has been simply to purchase their carsharing rivals. Carsharing businesses are a slightly different business model to the standard vehicle rental businesses but both involve owning (or leasing) and maintaining vehicle fleets available for hire and generally for use within a specific territory. From the consumers' perspective, carsharing and conventional vehicle rental are quite different services, but from a corporate perspective, they are basically the same kind of business with somewhat different customer bases and types of service. Vehicle hire has always been a highly competitive business and although the established firms would not have welcomed the entry of new carsharing businesses, the extent of the disruption does not appear to have been threatening to the viability of the established vehicle hire businesses around the world, at least until now. In part, this may be because carsharing businesses have a customer base that differs somewhat from the established vehicle rental businesses and the locations and territories serviced tended also differ, so a degree of co-existence is commercially tolerable. Complaints over commercial bikesharing tend to be largely from a minority of users of the service with typical service-related consumer grumbles; there seem to be very few substantive complaints presumably, because there was little widespread disruption to vested interests in the transport sector. Bikeshare schemes receiving public subsidy, however, have been criticized by those opposed to such support for cycling.

Ridesharing schemes, by way of contrast, have proved far more disruptive, controversial and contentious than carsharing businesses. An obvious difference stems from the fact that, generally, vehicle rental occurs in an open market and ridesharing impinges on the interests of those involved in a closed market. In other words, most places around the world have few special restrictions on firms wishing to found new vehicle rental businesses, whereas there are typically many restrictions imposed by governments facing businesses wishing to establish a new taxi service. These restrictions typically tend to involve two kinds of broad restraints: the necessity of a licence to operate, the numbers of which are restricted; and the imposition of limits on service territories. Operating licences can take a variety of forms, including annual permits and tradable permits; usually there are also forms of driver training and special driver permits required. Taxi businesses take a variety of corporate forms and drivers may be contracted employees, casual employees, business partners or sole owner-operators. There are also additional obligations and restrictions governing taxi businesses, notably government controls over fares and pricing, levels of service standard and driver behaviour, special safety obligations, service monitoring and customer complaint provisions. In short, taxi businesses are highly regulated in comparison to ridesharing enterprises. A major aspect of the disruptive character of ridesharing is that the ridesharing consumer is largely indistinguishable from a taxi customer. For taxi companies, nearly every ride-sharing consumer is a lost customer; commercial ride-sharing and taxi companies are in direct competition within the same market.

Commercial ridesharing, as described above, has begun as an essentially unregulated business operating directly in a highly regulated and closed market thereby upsetting both government regulators and existing taxi businesses and taxi operators/drivers. Taxi companies complain that they have invested heavily in their businesses in order to satisfy the demands of regulators and must operate within strict limits, many of which were imposed by legislation in order to protect customer interests by setting fares, ensuring comfort and safety, eliminating discrimination and providing order to the business. They point out that private car drivers taking passengers for a fee are essentially breaking the laws that govern taxi businesses and furthermore, are setting fees free of the financial burdens carried by legitimate taxi businesses. Licenced taxi drivers' complaints against ridesharing are based around the treats to their livelihoods and that rideshare providers do not have to meet the same standards regarding training, official monitoring, safety and comfort responsibilities and the like.

In the US, Uber Technologies is the largest and best known of these companies (its nearest competitor, Lyft, is about one-seventh of the size). Uber's approach to becoming established has been to start operations in a location regardless of the state of regulations and then in the face of official opposition to then seek recourse through the legal system. This has resulted in a welter of government responses, such as fines, being sued and being instructed to cease operations. To date, the company has been largely successful in overturning the restrictions imposed upon it through intense lobbying efforts within US state legislatures, although many states are continuing with various bills and legal actions to bring Uber under tighter regulatory control.

Rideshare customers have experienced the kinds of market abuse that can be expected from an unregulated service. For example, Uber customers have complained of 'price surging', referring to higher charges during times of peak demand, stating that the price information they had received before accepting a ride was that applying to regular trips; customer complaints resulted in the US Better Business Bureau giving the company its lowest ranking in 2014, an 'F' (Horowitz 2014). Uber's stance on privacy has also been questioned after suggestions that it was using information on its customers in malfeasant ways to discredit journalists, critics and competitors were brought to light (Reilly 2014). Unlike the anonymity of taxi use, commercial ride sharing operates on user accounts and in all likelihood has the capacity to track individuals through the applications on their mobile phones. In the US, a controversy has arisen over Uber's privacy policies. There have also been failings in Uber's screening of drivers; following a charge of rape by an Uber driver in New Delhi in 2014, that city banned all Uber services when it was revealed that the accused had been previously arrested for similar crimes (Menon 2014). Two Uber drivers have been charged in Chicago for attacks on passengers in late 2014 and early 2015. Other places have also partially or wholly banned Uber, including Spain, France and Thailand. Uber drivers, as a non-unionised, self-employed and casual workforce are vulnerable to the pricing policies of the corporation. Many of the strongest reactions have come from taxi drivers; in June 2015, there was a riot in Paris by taxi drivers during a national protest against Uber.

Essentially, the claims of the taxi industry have merit; they have a sanctioned monopoly to provide a public service and must meet regulatory guidelines in providing those services. In effect, the private providers of ride sharing are challenging the validity of the closed market on the grounds that it need not be closed. What is perhaps ironical is that it is the failings of the taxi industry that have made commercial ride-sharing so attractive and successful in such a short time. Rather than using contemporary communication technologies to help resolve these traditional problems with taxis, the industry is being overtaken by operators who have found solutions to these problems and are disrupting the established business model. More fundamentally, what the rideshare companies have also exposed is the rationale for taxi businesses being sanctioned to participate in a closed market by governments, and in a number of ways, a questioning of this rationale was overdue.

5 Eco-Communalism

Living communally or adopting a cooperative or self-organising approach for producing, consuming and distributing goods and services has roots in specific political philosophies, ideas and ideals. Undertaking such arrangements as a strategy for environmentally sustainable living created a new type of communal living, or at least, a new rationale for doing so. Community ownership of transport can be seen as fitting in to this form of *eco-communalism*. Knowing something of the origins of these ideas provides insights into this aspect of community-owned transport and places the concept into its political and philosophical context.

Ecological Communities

Ever since the advent of the Industrial Revolution there has been a movement and belief that many of the worst social and environmental effects of industrial society have a remedy in the restoration of communities with closer social ties and relationships to local ecology and within social groups. Community, in this context, refers not only to social form, namely smaller-scale social groupings living cooperatively within a defined physical territory, but also to idealised social relations that emphasise cooperative behaviour and mutual obligations. Environmental advocates and scholars have promoted the virtues of such communities offering a viable alternative to the usual aspects of industrial society that has given rise to multiple global and local environmental crises. Earnest Callenbach's famous 1975 novel, *Ecotopia*, sets the tone with its eco-utopian vision of cities where cars are banned and people and goods are moved by trains and buses. This chapter considers the issue of such eco-communities from a philosophical perspective and contends that these key ideas are the foundation of community-owned transport. Eco-communal living also represents a broad range of practices, with many hundreds of ecovillages around the world. However, as discussed in this chapter, the role of transport in the world's ecovillages is typically a neglected or minor theme, for a variety of reasons.

Eco-Communalism as an Environmental Discourse

Eco-communalism refers to the (self-conscious) belief in, and practice of, environmentally sustainable living at the local scale through self-sufficiency. Its central themes are self-reliance, human scale, decentralisation, community cooperation and environmentally appropriate technologies. Although decentralisation and non-urban living are strong themes in eco-communalism, these are necessarily of less interest and relevance to this volume; however, they are covered in this account. As Karen Lifkin (2014) states, ecovillages have diverse lineages, including in ideals of self-sufficiency and spiritual inquiry, such as monasteries and ashrams, the 1960s and 1970s social movements, 'back-to-the-land' and cooperative housing movements in developed nations of the 1960s, and the participatory development and appropriate technology movements in developing countries. Some consider ecovillages to be a wider social change that covers not only alternative lifestyle and environmental movements, but also a range of alternative banking and finance institutions, international development projects and permaculture. This diverse perspective tends to confuse matters and diffuses some of the specific features of existing eco-community settlements.

As an environmental philosophy, eco-communalism covers an overlapping space with utopian and environmental anarchism; its followers reject materialistic and instrumental valuations of Nature and seek social development occurring in ways that adapt to natural conditions. As Carter (Carter, N. 2001, p. 71) states: 'Anarchism is, in many respects, the political tradition closest to an ecological perspective'. He goes on to summarise:

> Core green principles like decentralisation, participatory democracy and social justice are central features of the anarchist tradition, and many greens have inherited the anarchist distrust of the state. Anarchists have also helped the praxis of green politics by advocating grassroots democracy, extra-parliamentary activities and direct action.
>
> (Carter, N. 2001, p. 71)

Eco-communalism qualifies as a social movement of sorts, as its followers express a collective belief and act to pursue it, it has shared interests and it seeks change outside established institutions (Martell 1994). Although eco-communalism has political motivations and actions, it is perhaps more clearly understood in sociological terms involving a new social group (i.e. it concerns structural change for a new economic structure) and cultural change seeking changes in social values.

Dryzek (1997) places eco-communalism within green romantic discourse. Intellectually, there is a rich heritage behind the belief, drawing broadly from social progressive and emancipatory thinkers, seventeenth-century Romanticism, communitarianism and a variety of environmental philosophical threads. Part of the roots of eco-communalism reach back to what Guha (2000) describes as the first phase of environmentalism in the 'back to the land' movement built on a critique of industrialisation that rejected the Industrial Revolution. Carter (Carter, N. 2001) considers eco-communalism a variant of eco-anarchism (as does Eckersley 1992), and links it to a belief in the need for a closer relationship between communities and the local environment and that 'the state is intrinsically inimical to green ecological and social values'. Eckersley (1992) divides eco-anarchism into eco-communalism and social ecology; this is a helpful clarification, as we are more concerned here with eco-communalism, although social ecology is summarised below.

Another strand of this thought is bioregionalism. Here, the idea is that geographical areas with homogenous features, such as climate, geomorphology and vegetation, provide the foundation for social organisation. Several authors identify eco-communalism with bioregional thinking (e.g. Pepper 1993), of which Sale (1985) offered four features: (1) *Scale*: regional and community, (2) *Economy*: conservation, stability, self-sufficiency and cooperation, (3) *Policy*: decentralisation, complementarity and diversity and (4) *Society*: symbiosis, evolution and division. While some aspects of Sale's bioregionalism draw on aspects of anarchism, other aspects are closely tied to environmental thinking. Regarding the former, Sale offers that bioregionalism concerns developing the self in ways not determined by the state and bureaucracy and in developing a region so as to support a community. Regionalism's environmental links, following Sale, are an intimate knowledge of the land and developing an understanding of the traditional knowledge and culture of the area.

More recent commentary has cast eco-communalism as a rejection of modernity, albeit that the major themes in such work strongly resembles those of the preceding description. From this perspective, modernity has been a force for what is rather inelegantly described as the 'de-localisation' of communities. Transport has been a major force in this process, with mechanised and low-cost transport acting to disperse the distribution of productive and consumptive activities. Post-industrial processes have been enhanced by globalisation and neoliberalism, leading to the loss of local services (such as schools, health centres, banks and government services) and the rise of more centralised and larger providers.

Eco-Communalism as a Social Movement

As practiced, eco-communities represent a new social movement, rather than a traditional social movement seeking participation in established political systems, and it can be instructive to view these social experiments in communal living in this light. They qualify as a new social movement by being located in civil society rather than in the polity, are concerned with values and lifestyles rather than seeking political integration and economic rights, and as a group are typically more grass roots-oriented and involve networks and informal associations and avoid formal and hierarchical structures. So although eco-communalism has distinct political and philosophical roots, it does not constitute a political movement in a conventional sense and, as discussed below, this constitutes both a strength and a weakness.

Whether or not eco-communalism amounts to a political movement, however, can be contested. Indeed, whether eco-communalism constitutes a movement at all is questionable. Using a tight understanding of a political movement, eco-communalism probably does not qualify as it is without any of the common and overarching principles and goals that would give such a movement a clear identity. Under the broad diversity of ecovillage projects are many different political identities and indeed, it may well be that a good number of the inhabitants of ecovillages are not necessarily engaged in formal political activities and projects. For some participants, ecovillage living is a deliberate retreat from many aspects of conventional industrial society, including its formal political practices and debates. More critically, eco-communalism is not a directly disruptive movement in a political sense; many of its followers may oppose capitalist economics, centralised political authority in state institutions and other institutions, but are not organised in ways to subvert or directly oppose such institutions and practices. As described above, it is a sense of shared values, outlooks and personal aspirations that unites those committed to eco-communalism rather than any overt political programme.

Ecovillages attract the interest of political theorists and scholars involved in green politics not because they constitute a political movement per se, but because they represent political activity in a broader sense and their activities have political implications, especially for environmental politics. There will be no revolution for re-shaping nation states for environmental protection as a result of the ecovillage movement, but with equal confidence it can be said that any major ecological reform of existing political institutions will invariably draw on the lessons and insights drawn from the experiences of ecovillages that have sought to put environmental ideals into practice. Indeed, it might be argued that many current environmental initiatives, especially at the municipal level, have their roots in the experiments and initiatives of ecovillages. Municipal recycling and re-use centres, local financing for environmental projects, the use of permaculture and other local production systems and support for community food-growing might exemplify the influence of ecovillages. If this is the case, then ecovillages represent a cultural influence, at least in the developed world of the West, rather than a political influence.

Environmental Rationale for Strengthening Neighbourhoods and Communities

Eco-communities span a range of community types, including rural villages, new urban developments, urban renewal projects and townships. It has long been held that the local- and community-scale initiatives and practices offer special advantages in living sustainably. Most aspects of the environmental rationale for eco-communities are well known and without controversy, although the monitoring and investigation of the environmental performance of eco-communities remains patchy. Complicating the analysis of the claims for eco-communities is that these communities attract individuals and households with pre-existing commitments to ecologically sustainable living, so there is a chicken-and-egg problem in distinguishing between the influence of the participants on practices and of the influences on practices mediated by the design and functions of the village on the inhabitants. In any event, these sites exemplify best practices in many aspects of ecologically sustainable living and have been highly influential models for other communities and in public policy formulation, especially by local governments and municipalities.

Transport has typically been a smaller theme in most eco-communities, where the higher priorities have been given to sustainable building and architecture, residential energy supply, food production, water supply and wastewater treatment, materials recycling and re-use, and institutional and local political interests, covering community management, group decision-making, leadership and related matters. Nonetheless, some eco-communities have addressed transport issues and there can be many indirect transport benefits arising from other aspects of eco-community living, particularly with travel demand reduction.

Eco-communities can do a number of things to reduce motor vehicle use and thereby lower transport greenhouse gas emissions, such as reducing the need for travel and finding alternatives to car use. Local production, such as of food, reduces the need to travel for food shopping. Local employment too, has a similar effect in reducing travel and shorter trips allow for a greater choice in transport mode, notably using public and active transport. Concentration of local services, including education, health, leisure and retailing, can facilitate cycling and walking. Reducing or eliminated reliance on cars can be achieved by creating mobility alternatives that offer greater diversity in transport choices. Creating healthier local environments can be achieved through sustainable transport practices and achieving lower levels of pollutants and noise pollution, less time used in travelling, lower expenditures on transport and lower levels of stress and anxiety. Greater active transport confers health and wellbeing benefits and reduces the mortality and morbidity effects now associated with sedentary lifestyles that in developed nations afflict both urban and rural communities. Safer streets may be produced through a number of design measures and also by reducing the overall use of motor vehicles for transport. Street design can lower vehicle speeds, restrict vehicle movement, allow streets to be used for multiple purposes as places of open space and make them into an effective public realm. Local accessibility to transport is improved by such measures and this generates greater equity in mobility compared with transport systems dominated by private car ownership. Increased self-determination in transport can also be associated with more decentralised systems by giving users greater control over transport systems. Smaller communities can adopt models of direct democracy, using trust and mutual obligations to effectively manage individual and collective resource use and waste production. Many of the attributes that public policies and urban planning are now seeking using environmentally sustainable transport and urban planning have already been established in ecovillages over several decades.

Eco-Communalism as a Green Philosophy

Unlike some environmental discourses, there is no distinct philosophy behind eco-communalism as it embraces several different themes. As Pepper (1993), Eckersley (1992) and others have shown, the relationships and differences between these different aspects of ecological thinking are highly complex. It almost goes without saying that advocates of these ideas are strongly critical of each other, as in the disagreements between advocates of deep ecology and advocates of social ecology.

Political philosophies are usually arranged along the left–right axis, but some regard this dichotomy as an outmoded device. Despite contemporary criticism, this difference remains widespread in both popular culture and political scholarship, perhaps because of the sheer weight of history and because these simple differences often have greater explanatory power than more sophisticated alternatives. Environmental philosophies have ties to classic political philosophies and can be catalogued as being closer to the left or right wing, but there are also aspects of green philosophy that confound the left and right division. This ambivalence plays out in part in the ideologies of Green political parties, for example, which have been identified as new social movements that reject the traditional party allegiances and policies. Mapping out the rough terrain of environmental philosophy, or 'ecologism' (a term used by Dobson [2000]), denotes the field of green political ideology as a collection of beliefs. In terms of its political preferences regarding economic systems, ecologism is fairly clear in what it does not support. Ecologism, in general, rejects unfettered capitalism and seeks its reform, just as it rejects unrestrained materialism; however, it does not endorse a centralised command economy under state control (i.e. communism). In this volume, we are interested in two aspects of ecologism that occupy contrasting positions: eco-communalists and green reformers, between which, there are many ideological differences.

Utopian Socialism and Anarchism

Utopian socialism lies at the root of much eco-centric philosophy (Pepper 1996). Prominent in the long sweep of early proponents of egalitarian and communalistic societies (a movement that can traced through to pre-industrial times), was the more familiar utopian socialism of the eighteenth and early-nineteenth centuries, such as developed by the early utopian socialists Claude Henri de Saint-Simon, Charles Fourier and Robert Owen (Pepper 1996). Pepper (1996) states that these thinkers favoured decentralising power to small communities, rejected social hierarchy and leaders, and expressed scepticism over private property and industrialism. Decentralised, self-sufficient and autarkic communities sought by utopian socialists match many of ideal communities sought under ecologism. These were also forms of socialism that sought closeness to the natural world, a peaceful transition to the future, benign science and technology and work related to human creativity (Pepper 1996). Followed by the anarchists of the nineteenth century, Pepper considered the anarcho-communist Peter Kropotkin to be closest to ecologism, with William Morris of the Victorian utopian socialists having the strongest interest in environmental interests. Throughout the twentieth century there were efforts to try and create utopian socialist settlements, including back-to-the-land movements.

Where these sort of utopian socialists depart company from Marxists is their willingness to articulate the society they sought, whereas Marxists, following Karl Marx, believed that the design of future societies was the business of those societies (Pepper 1996). In practice, Marxists and anarchists alike had visions for the future (which attracted the critiques of liberals) and an intolerance for varying views, factors that invariably pulled their visions towards totalitarian outcomes. True to their ideals, utopian socialists adopted a utopian approach towards political processes, placing the role of moral aspirations at the core of their programme; changed personal values would precede political action, which in turn, precedes overturning capitalism. Theirs was a battle against existing religious and political ideologies, with conventional political processes being rejected. Establishing communities to live out these ideals saw efforts to put utopian socialism into practice; many ecocentric ideals and practices become part of new intentional communities, beginning in the seventeenth century and continuing into the present day. In England, for example, it is noteworthy that a number of these communities were founded in the seventeenth and eighteenth centuries as a consequence of the grazing lands commons enclosure underway in those times (Wall 2014). However, the early utopian socialists were, almost without exception, intellectuals and members of the middle class, rejecting what early industrialisation had visited on the working classes (Pepper 1996).

Much of historical anarchism can be found in contemporary ecologism (Pepper 1996, Eckersley 1992). Anarchism endorses self-government and opposes conventional government and the state, promoting consensus in decision-making collectives and communes. States, under anarchism, are replaced by free associations and group cooperation; it is through communal associations that individuals are able to live naturally. Different forms of anarchism include anarcho-anarchism (community-based associations) and anarcho-syndicalism (workplace-based associations). As Pepper (1996) states, environmentalists align with anarchists in things favoured, such as individualism or collectivism, egalitarianism, volunteerism, federalism, decentralism, ruralism and altruism/mutual aid; and things rejected, including capitalism, giantism, hierarchies, centralism, urbanism and competitiveness. Anarchist theory stressed the idea of a natural society (as in William Godwin's *Political Justice*), free of the hierarchies created by states and favoured human traits of mutual aid (as in Kropotkin's *Mutual Aid*), as opposed to Darwinian competition (Pepper 1996).

Individuals achieve the highest quality of life, in this philosophy, when they have the greatest autonomy in decision-making about their lives in small-scale societies. Anarchists were divided over the whether the individual or the community provided the best setting for individual decision-making, with Kropotkin advocating the latter (Pepper 1996).

Communalism and Individualism

Cooperation is taken as a natural human trait by anarchist thinkers, a virtue that is corrupted by inappropriate institutions based on hierarchies, as opposed to classic liberalism, that holds human self-interest as an inherent trait that must be restrained, such as through government and a social contract. As Baxter (1999) summarises, communitarianism stands in opposition to classic liberalism in its understanding of individual autonomy. Four features define this position: 1) Values and purposes for individuals are defined socially, not individually; 2) Such values are derived from concrete social contexts not from universal sources; 3) Each context, with its own traditions and values, endorses a 'specific vision of the good life'; and 4) There can be no neutral position for governments in defining the good life, rather its task is maintaining and furthering each society's own specific view of what constitutes the good life.

Community is broad concept in the social sciences, although in contemporary industrial cities, as Luke (1997, p. 183) observes, there is not much 'community' present, it having 'very little popular form or independent social content'. Although his comments refer to the US, Luke identifies the effect of high 'personal, economic, geographic, and social mobility', so that as a consequence of the divisions between workplace, residence, production and consumption, and between identity and interests, community becomes 'thinned', as applicable to most OECD cities. As a result, he argues, interests are divided and common historical consciousness, shared beliefs and ecological responsibility are lost. Eco-communalism stands therefore, as an ideal and practice contrary to both contemporary urban life and to those economic and political interests exercising political power through state governments and capitalist markets. In other words, eco-communitarianism's promotion of a deeper practice of community in modern society may be as revolutionary as seeking to restrain economic activity within ecological limits.

Centralised and Decentralised Living

If economic growth and technological development are to be controlled, then society needs to be organised to achieve such a monumental shift towards an eco-centric future and this brings the reforming capabilities of democracies into question. Casting such fundamental ecological problems into the language of crises elicited responses that included the far ends of the political spectrum, namely centralised and authoritarian solutions (e.g. such as those of William Ophuls, Garrett Hardin and Robert Heibroner) and decentralised and participatory solutions, based in eco-socialism. Given that conventional state and corporate institutions are rejected by eco-community supporters, the desirability of an enlarged authoritarian state, the validity of its capacity for environmental management and the sustainability of such institutions would never be accepted by this group.

Part of the refutation of industrial society was articulated in an environmental classic, E.F. Schumacher's (1973) *Small is Beautiful*, that rejected economic growth and industrial technologies for the harms they caused society and ecology. From this foundation were a range of solutions, including calls for eco-technology, decentralisation as a way to restrain consumption, restoration of social justice and discovery of ways for living within ecological limits. It follows that such goals should all be realized in an idealised form of society, the ecovillage.

A well-known form of this advocacy draws on references to monastic life, such as Rudolf Bahro's (1986) concept of 'ecofundamentalism'. While monasticism conjures images of denial and an orthodoxy of faith, within ecologism was a theme of creating communities with a culture and social order based around a harmonious relationship with the local, natural environment, so-called 'liberated zones'.

Decentralised ecovillage lifestyles differ from the paradigm of centralised urban living as exemplified in modern industrial states in a number of ways. Politically, the central aspect of decentralisation is the eco-communities' systems of government and management based on locally self-determined institutions, processes and activities. Being nested within larger states with their layers of governance institutions, the self-governing of ecovillages does not lessen the burden of governance on ecovillagers, rather increases the degree of governance but for the benefit of direct citizens involved in local decision-making. Further, because of the collective character of ecovillages, the burdens of ecovillage responsibilities fall more equitably than would be expected in conventional urban living where the disparities in income, status and opportunities are broader and are part of institutions that reinforce inequitable distributions of political influence.

Decentralisation is also exhibited in the access to and use of utilities for services, such as household energy, water and wastewater, where ecovillages use local and collective systems, rather than those of large state or corporate utilities. These technologies can be of local design to satisfy local requirements and to satisfy the need for achieving low environmental and social impacts. Ecovillages have, therefore, greater control over the design, use and management of such systems and are not subject to the preferences of distant and largely unaccountable entities that control centralised utility services. Ecovillages can produce their own foodstuffs and be self-sufficient and may also provide for other material needs, such as timber, natural fibres and other material resources. Such independence allows ecovillages to reduce largely or perhaps entirely, their reliance on outside markets and the consumer economy. Other services that might be normally produced by governments or private firms, such as schooling and education, health services, recreation facilities, libraries and transport can also be provided locally on a collective basis, thereby achieving independence in the means of production. Further independence from the wider economy and its controlling institutions can be achieved by locally controlled retailing and by creating independent currency systems and barter schemes. These attributes form part of an intentional strategy to separate key aspects of ecovillage life from the centralising forces of governments and the wider economy.

Industrialism and Eco-Technology

Environmentalism as a whole has largely rejected industrialisation with its twin engines of economic growth and technological development. Yet ecological thinking can be arranged along a spectrum according to the place of technology in addressing the environmental crisis (O'Riordan 1981). Technological optimism is marked by the view that technologies of appropriate design and operation can be directed towards environmental reform, the most extreme form of which is technocentrism, which holds that technological change is both imperative and sufficient to address environmental problems without (significant) social and economic change or disruption. There is an on-going commitment to economic growth in technocentrism and, by implication, resource depletion and ecological loss are but further challenges.

At the other end of the spectrum are more eco-centric views, marked by scepticism, if not rejection, of technological solutions to environmental problems. Resource and ecosystem protection are given the highest priority, with resource consumption and waste production to be at environmentally sustainable levels. Not only is mass consumption rejected but, more widely, materialism is identified as part of the ecological problem, so that solutions are sought in realms such as participation, cooperation, personal responsibility and education. Humility towards the natural world is emphasised, so that expertise in environmental management and confidence in technological solutions are regarded with doubt.

Between these poles lies a range of more positive views towards the potential of eco-technologies. How the major environmental philosophies regard technology is a major defining feature, but as Carter (Carter, N. 2001) and others have pointed out, attitudes towards technology do not provide a clear left wing versus right wing differentiation. Reforming industrial society to meet environmental sustainability goals invariably requires a technological transformation and within ecologism there is a general consensus that technological solutions are required. Where the major political and philosophical differences emerge is over the extent to which technology can serve this end, particularly in relation to the development needs of developing nations. One of the broader debate's major divisions is the extent to which technologies can reduce environmental harms, arguments that in the 1970s gave rise to identifying technologies most suited to environmental and socially progressive goals (Schumacher 1973) – the appropriate technologies. Such appropriate technologies are characterised by these features: low negative social and environmental impacts, relatively small scale, being suited to a local context (such as geography, culture and politics) and controlled (and possibly maintained) by local citizens or communities. It is these forms of technology that appear most suited to the technology needs of ecovillages.

Bookchin's Social Ecology

Although here we are most interested in the eco-communalism aspects of eco-anarchism, there are aspects of the other major aspect of eco-anarchism – social ecology – that are worth considering. Prominent in this field and the individual most usually associated with social ecology is Murray Bookchin. At the core of Bookchin's beliefs is the link between the ecological crisis and social hierarchies, so that his ecological solutions are based on emancipation. This argument identifies the core of environmental problems as domination within society, drawing on Kropotkin's idea that Nature is egalitarian and Bookchin (1980) finds that ecosystems are without hierarchy and its natural systems are interdependent.

From this foundation, a system of environmental ethics is devised: 'If nature provided the ground for an ethics that has an objective ancestry in evolution's thrust toward freedom, selfhood and reason, so too nature provides the ground for the emergence of society' (Bookchin 1986, p. 16).

Bookchin applies this principle to society, finding that people are naturally cooperative and function best in decentralised (i.e. anarchic) groups and draws on traditional societies as a reference point. Such societies applied the same perspective to their relations with the natural world, so that they neither sought to dominate Nature nor to be subject to its domination. From that point, states Bookchin, the movement towards hierarchies in society based on age, gender, religion and other factors created societies that began to seek domination over Nature. Because Nature is viewed separately from society and regarded as cruel, competitive and hostile and as humanity's opposite, forming a barrier to social freedom and self-realisation, a dualism was created between society and Nature, argues Bookchin. Consequently, society sought to tame and exploit Nature for social ends through 'reason, science, and technology', regardless of the costs and damage to the environment, causing 'a fragmentation of humanity into hierarchies, classes, state institutions, gender, and ethnic divisions' (Bookchin 1986, p. 53).

Bookchin's social ecology seeks a break with such values, thinking and institutions. As he states:

> Social ecology is, first of all, a sensibility that includes not only a critique of hierarchy and domination, but a reconstructive outlook that advances participatory concept of 'otherness' and a new appreciation of differentiation as a social and biological desideratum.
>
> (Bookchin 1986, p. 25)

Resulting from this systemic domination within hierarchal societies, Bookchin claims, are a series of dualisms in society that are inimical to addressing environmental problems. These dualisms include the higher status of intellectual work over physical work and work over pleasure.

Critiques and Limitations

Critiques of eco-communalism come from conservatives and, perhaps with the usual trenchancy, from the left wing, especially from followers of other variants of green politics (although several specific criticisms are common to opposing groups). Many criticisms of eco-communalism are drawn from general critiques of anarchism. A frequent criticism is that eco-communalism romanticises pre-modern and traditional societies (identifying them as organic, with close social ties and of altruistic character) and draws false dichotomies against contemporary society when claiming that these are selfish, materialistic, regimented, alienating and bureaucratic. Supporters of the status quo often contend that modern states and their political and social institutions underpin virtuous aspects of modern life, as made evident by the widespread advances in the quality of life. If the state is dismantled, argue the critics of green anarchism, there is no guarantee that voluntary cooperation will work and that society will not regress to a more primitive, divided and iniquitous condition.

Bioregionalism has been criticised for a basic incompatibility between coherent bioregions and actual human settlements (particularly the ecological footprint from which natural resources and services are drawn) and the political and social boundaries used for government and administration. Essentially, although traditional societies were within, more or less, bioregions, industrial societies have crossed and transcended such limits and biological restraints. As Alexander (1990) stated, our consciousness of regions is shaped by cultural factors, including language, religion and social ties, rather than through strict biophysical geography. That bioregions might be the basis for governance also stands in opposition to the real-world composition of the populations within bioregions and the problem of governance across multiple bioregions. Bioregionalism appears poorly equipped, point out its critics, to deal with global environmental problems (Eckersley 1992). To this is added the problem that simply possessing a bioregional awareness is hardly sufficient to ensure that governance results in environmental protection or indeed, adopts a cooperative approach.

Bookchin's social ecology views have attracted considerable attention and criticism (e.g. Eckersley 1992, Light 1998). Amongst these criticisms are that Bookchin's ideas are an anthropomorphising of ecological phenomena according to a pre-existing set of political beliefs and these phenomena have been equally interpreted to reach diametrically opposed conclusions (such as in the case of social Darwinism). Social ecology's belief in the non-hierarchal societies of the pre-literate era has proved to be a shaky foundation, with critics finding many exemplars of deeply stratified and hierarchal societies living harmoniously in their natural settings. Eckersley (1992) pointed to the opposite problem, namely that an egalitarian society may still exploit Nature. A series of famous exchanges between Bookchin and deep ecology supporters highlighted Bookchin's antipathy towards the prospects for reform through personal enlightenment; further, he thought much of the deep ecology programme to be misanthropic, such as its support of population control. Despite such differences, social ecology and deep ecology share a view of the state as inimical to ecology and social interests.

Utopian ambitions have attracted several consistent themes of criticism based both on political theory and efforts to create such communities and these have all been made against eco-communalism, namely the charges that eco-communities are conservative, romantic and totalitarian. Critics point to the ways in which traditional societies fail to meet the requirements of contemporary society, pointing to prevalent instances of violence, social inequality and oppression. Returning to such pre-modern conditions, they argue, is not only inimical to any prospect of utopian society but is essentially regressive. Linked to this critique is the charge that eco-communities have adopted a romantic vision of earlier societies, taking from the historical fabric those selective attributes that appeal and ignoring the wider components of those societies, or simply imagining social attributes that never existed.

Another line of criticism is that in rejecting established social institutions for social order and conventional democratic politics, eco-communities are destined to become totalitarian. Certainly, for those communities founded on charismatic leaders, such a charge has some merit, whereas others have created effective collectivist institutions of their own to satisfy contemporary demands of social justice. Overlapping this critique is the more focused claim that ecovillages are little concerned with democracy. That many ecovillages express support and have a role for spirituality is taken by some to be inimical to democracy, for while state authority might be rejected, religious authority can come to assume this role and be, therefore, contrary to the interests and practices of democracy.

In sum, these criticisms of eco-communalism have it opposed to modernity, a position deemed to be misguided and futile. Anthony Giddens (1990), for example, in *The Consequences of Modernity*, warns of 'romanticised views' of communities in pre-modern times, including those in cities, where people were 'relatively immobile and isolated'. Taylor (2000) is sceptical of the ecological potential for ecovillages as they run counter to modernity and fail to engage with global environmental problems. However, he concedes that: 1) There are local environmental benefits to be gained through local initiatives, even if local lifestyles are unaltered and 2) There may be environmental and 'quality of life' benefits of resisting some forms of 'de-localising' modernity. Notwithstanding this concession, Taylor considers that ecovillages have few prospects for wider success as 'the continued predominance of the non-localizing tendencies of modernity diminishes the chances of this project being successfully realized' (2000, p. 27).

Eco-communalism and ecovillages, almost paradoxically in view of the preceding comments, are also criticised for their absence of an overt political agenda to address the major economic, social and cultural institutions at the basis of the environmental crisis. Yet Lifkin offers a different interpretation of this outlook and wrote in praise of ecovillages:

> Unlike the larger environmental movement, ecovillages are not primarily concerned with protesting against the state and industry lassitude. Rather than waiting for the revolution, they are *prefiguring* a viable future by creating parallel structures for self-governance in the midst of the prevailing social order.
>
> (Lifkin 2012, p. 131)

Virtues in the eyes of Lifkin, Benton (1995) and others, are failings to those endorsing a revolution to create an ecologically informed social order. By withdrawing from mainstream industrial society, the ecovillage critics argue that ecovillagers are unable to exert wider influence and have no capacity to undermine or reform the existing institutions of states and corporations.

Part of the scepticism of the defenders of modernity towards eco-communities should be kept in perspective and at least respect the claims being made for eco-communities. Revitalisation of neighbourhood scale activity for environmental goals need not be in conflict with global environmental goals and the necessity to respond with national and state public policies to contribute to global efforts;

presumably, few ecovillage advocates would oppose such actions. Indeed, despite the enthusiasm of ecovillage advocacy, there are not many claims that ecovillages obviate other kinds of social and economic change for environmental protection. Partly what we also see in this debate is the differences in environmental discourses, so it is natural that followers of ecological modernisation would find fault with romantic visions of many ecovillage advocates.

Garden Cities, New Towns and Ecotowns

City planners and designers also responded to the failings of urban growth under the Industrial Revolution and some of their responses were rooted in the same political values as those of contemporary socialists and bound in with social reform movements. Over time, the perceptions of these urban failings has evolved, for as Hall (1988) observed, urban planners reflect the ideals and social concerns of their era, from the need to address the squalor and moral turpitude of the industrial slums, to the modernist ideals of progress after WWII to the need for ecologically sustainable cities in our own era of ecology with the rise of the eco-city, a term attributed to Richard Register (Register 1987).

Perhaps the most famous of the early responses is Englishman Ebenezer Howard's garden cities concept promulgated in the late nineteenth century. This design featured cities of 32,000 inhabitants laid out concentrically and clustered with other garden cities around a major urban conurbation of 250,000 population, all joined by railways and roadways. Urban functions – residential, work, civic activity and agricultural land uses – were separated and rationally organised; open space was a feature of the layout, with houses having private gardens and access to allotments. These ideas were taken up in the UK (as Letchworth Garden City and Welwyn Garden City) and were influential in many US-city developments; it was also applied in Argentina, Australia, Brazil and elsewhere, often in colonial and post-colonial settings, but also within several European cities (Hall 1988). Garden cities established much of the principles of the UK's new towns movement of the post-war expansion that sought to both replace the poorly planned and overcrowded older cities and industrial towns with modern and self-sufficient towns. Similarly, other nations also built versions of new towns through the mid-century era, although the new town movement was more generally progressive and modernist than socialist in its political framing (except for those nations with centrally planned economies).

Success in realising the ideals of the garden city and new town movements was somewhat circumscribed by concessions made to the original designs and the morphing of the garden city into what is sometimes called the 'garden suburb' that came to dominate the suburban boom of the post-war period that has continued into the present. Politically, the garden city developments rarely come to exemplify any politically radical aspirations nor were the planners and developers involved notably democratic in exercising their professional duties. This is not to attribute the failings of the dispersed city and car-based model of suburban development to the garden city movement (although some critics have done so).

Many competing visions of the eco-city have arisen since the age of ecology in the 1960s and 70s and as the concept has become more popular, the range of its meanings and applications has broadened and become diffused to the point of ambiguity. Further, the new urbanism movement of the 1990s had strong elements of design for environmental sustainability (e.g. Calthorpe 1993). Eco-cities are now a global phenomenon, a mainstream activity at all scales with a great diversity of practices and approaches. While garden cities and new towns were designs for new cities, eco-cities embrace models for new towns and retrofitting of existing cities, covering maturing economies of the developing world, European-style cities of Europe and wealthy Asian nations and US-style cities as found in North America and Australia. Further, the eco-city label has been retrospectively applied to cities that have not deliberately self-identified as eco-cities. As Rappoport (2014) describes, while the earlier visions for eco-cities were intended to be citizen-led and motivated by social equity and environmental protection, most projects (especially in Asia and the Middle East) are large state and public sector projects of an ambitious and technology-driven character. Much effort under the eco-city banner follows the precept of ecological modernisation (see Hajer 1995), with technology and state-directed corporate activity playing a central role.

Transport, although not a prominent feature in garden cities or new towns, often takes a major role in eco-city designs and concepts, and occupies a central role in some visions (Register 1987, Kenworthy 2006, Newman *et al.* 2009). Eco-city designs for transport (and integrated land-use planning) draw on the established sustainable transport research using such themes as integration between the planning of land use and transport, low-carbon transport options, pedestrian and cycling facilities, mixed-use urban form, car-free development and transport and human health. Eco-city and new urbanism settlements can have more sustainable travel patterns, but such outcomes are not assured and a broader understanding of the explanatory reasons has yet to be developed. A number of large eco-city proposals and developments have, however, largely neglected the opportunities for sustainable transport. To date, the story of eco-cities is one of mixed and partial success, with some instances of outright failures against environmental objectives, proving once again that planner's models for Utopian urban life are difficult to implement.

Despite the limitations of the garden city and new towns movements and perverting of their goals and practices, nonetheless both provided antecedents of the current eco-city concepts and designs. Notably, eco-city designs have also usually dispensed with any visible manifestations of progressive or left-of-centre politics (or use such passive goals as the 'nurturing' of social justice); indeed, many seek opportunities identified within the practices of 'natural capitalism'. Kenworthy (2006), for example, acknowledges that the issues of poverty, inequality, politics and power are omitted from his list of key transport and planning dimensions. Caprotti (2015) makes the point that to neglect examining the industrial-economic system generating environmental costs 'risks focusing on symptoms rather than underlying conditions, on what is visible rather than on systemic (mis)functioning of political-economic life' (2015, p. 20). Implicated in this critique is the role of conventional models of urban planning as an elite, technical, authoritarian and static process. And, of course, there remains considerable resistance to eco-city initiatives from officials, corporations and homebuyers (Grant 2009). If we want to see what communities have done for themselves in urban design for environmental sustainability, while maintaining links to the values of eco-communalism, we need to look to the ecovillage phenomena.

Ecovillages: Putting Eco-Communalism into Practice

Putting environmental ideas into practice at the community scale has become, over time, perhaps the least heralded form of environmental activism, lacking the commercial potential of activism by professionals and professional associations, the newsworthy-ness of militant protest, the contemporary zeitgeist of household and personal enlightenment and the virtues of 'green' consumption. Our interest here is with community-scale responses to environmental challenges that go beyond the activism of awareness-raising and information dissemination. In the popular imagination, the presentation of eco-communalism in practice seems forever tied to the counter-culture from the so-called 'age of ecology' of the late 1960s and early 70s. This cliché has been an unfortunate and inaccurate picture of the diversity, seriousness and achievements of the community responses to the problems of ecological justice and environmental sustainability.

Ecovillages are eco-communalism in practice. They can be distinguished from conventional society through their common interest in ecological living, their formation of a distinctive community in which community relations are shaped by the members' common interest and that the community practices ecological living; i.e. these are intentional communities. A classic work in this field is Ted Trainer's *The Conserver Society* (Trainer 1995) that expresses a fundamental rationale for ecovillages thus

> No long-term vision for a sustainable world order can make sense if it does not focus on the people in a small region producing for themselves most of the things they need, from the land, labour, talent and capital of the region, and taking control over their local economic, social, community, political, cultural and ecological development.
>
> (Trainer 1995, p. 57)

Sociologically, ecovillages are communities with a shared identity and a common purpose that guides a set of practices that are meaningful to the community. These communities' participants are engaged in activities that align their personal beliefs with those of the community and are willing to abide by a set of formal rules and informal customs consistent with the communities' goals.

Lifkin (2012) captures the role of community in ecovillages in providing a wider and deeper response to environmental challenges than those centred on individualism and consumer preferences in mass markets:

> They are too small, both in material terms and as reflections of our humanity. This is because the creeping global crisis is not only a physical or political problem 'out there'. It is a crisis of meaning 'in here'. Therefore the truly effective alternatives will not only work pragmatically, they will also nurture within us a sense of deep belonging—to the planet and to each other. Given the scale and scope of the challenges ahead, our hunger for integrity can only be satisfied by the integration of ecology, economics, human solidarity, and inner meaning in our daily lives. This is precisely what ecovillages aim to do.
>
> (Lifkin 2012, pp. 131–132)

As the name evokes, ecovillages provide a co-joining of the ecological and the communal. Generally, the ecological side of ecological living has well-established principles that tend to the universal, such as the conservation of non-renewable materials, local production and self-sufficiency, minimising wastes, reduce–re-use–recycle and designing 'with' Nature. These principles are based on experiences and practices that have been generally supported through the empirical findings in the science of ecology and the experiences of self-sustaining societies. Social interactions in communal living do not come with the same sort of ecological knowledge base, for even within the more relatively uniform experience of modern industrial societies there are significant national, regional and cultural differences that shape social interactions. In addition to being places where communities work out how to live in accordance to these ecological principles, they are also 'experiments in radical democracy'. Human relationships are the more challenging and problematic aspects of ecovillages, providing both the source of disharmony and community discord but are also the wellspring of some of the most satisfying aspects of community living (Lifkin 2012).

It can be difficult to capture all that ecovillages represent, even if confined to the better documented cases in the developed world, because of the extent to which these are efforts to create new lifestyles that break with industrial modernity and also because so many ecovillages retain strong ties with, and rely on, features of industrial modernity, such as the use of modern renewable energy technologies, information sharing using the Internet and indeed, use of motor vehicles. These breaks with modernity aim at living in harmony with the environment and rejecting many of the tenets of contemporary developed world lifestyles including deliberate efforts to avoid mass consumption, to adopt less affluent living standards, and to minimise energy and natural resource consumption. Such decisions fly in the face of industrial modernity, but the means to achieving their environmental goals often employs the technologies of modernity and requires specialist technical skills and education. Socially and politically, ecovillages seek to develop their own local institutions, mechanisms for self-governance and community-based decision-making modes. Tenets of modernity, such as individualism and the economics of corporate capitalism, are rejected by ecovillages that typically endorse collectivism and cooperation.

As such, ecovillages represent some of the strongest tenets of democracy and display a greater commitment to its cause than is typically found in most modern nation states where public engagement in formal government process is usually very low. As such, ecovillages are a curious mixture of intentional lifestyles that reject exploitation and abuse of the environment and reach back to a selection of traditional village attributes, but generally embrace selected industrial technologies and govern themselves through a variety of democratic processes. In the developed world, it is incorrect to depict ecovillages as anti-modern and it is generally closer to the mark to consider them as living a form of high modernity lifestyle. Certainly, ecovillagers themselves consider their lifestyles as exemplars on which the future ecological conditions of the planet depend. As Barton (2005) points out, despite the array of contributors, perspectives and interests in ecovillage promotion, there is what he describes as a 'surprising coincidence of views' over what constitutes a sustainable neighbourhood: 'The rhetoric of sustainability talks of human-scale, mixed use and socially diverse neighbourhoods, providing residents with increased convenience and sense of identity, while at the same time reducing their ecological footprint' (2005, p. 11).

Before the age of the Internet, gathering information on the world's ecovillages was a difficult task. Now, however, there is a much larger array of information available on specific villages, electronic directories of online resources and a number of national and international ecovillage and associated organisations. According to the Global Ecovillage Network, there are about 400 designed eco-communities worldwide, but if developing world settlements are included, there are arguably another 15,000. Members of this network are diverse, including specific networks (such as Sarvodaya in Sri Lanka with 11,000 villages and Senegal's Colufifa network with 350 villages), eco-townships, rural collectives, urban renewal projects, education and learning centres.

There has been no shortage of utopian efforts to create new societies based on ecological principles. Lifkin (2014) has identified an impressive array of successful exemplars, focusing on 14 ecovillages. These represent the array of ecovillage experiences, varying in scale, economic models, organisation and governance, wealth and other factors. Analysed from four perspectives – ecology, economy, community and consciousness – Lifkin draws the analogy that each represents a window into the house of sustainability in which all windows must be maintained; the loss of any puts the ecovillage enterprise at risk.

Much of the discussion and investigation into ecovillages examines communities that were initiated, designed and functioning in the developed world, particularly Northern Europe, North America and the UK. As the membership of the Global Ecovillage Network shows, there are great many older communities in the developing world that satisfy the broad definitions of ecovillages. Although the concept of the ecovillage has been formally identified as alternative communities, there is now a wider recognition that ecovillages also have a rich and varied history that can be found in developing nations around the globe, pre-dating the coining of 'ecovillage', but beginning around the same time as their developed nation contemporaries. Developing world ecovillages stand in contrast to the intentional community ecovillages of the developed world, the former of which are traditional indigenous settlements with particular environmental concerns, such as preserving traditional cultures or preventing loss of local environmental values; creating sustainable livelihoods based on village enterprises is also an important motivation for ecovillage creation in developing nations.

Many, if not most efforts, have suffered from what Luke (1997, p. 170) wrote of the famous Paolo Soleri Cosanti project in Arizona, that set against its idealism were 'ideological fictions for stabilizing many myths in the present social order'. Aspects of Luke's specific critique could stand for a wider array of other efforts to create freestanding eco-communities and also evoke themes of political critique levelled at communitarianism. Schemes such as the Cosanti project are based on a 'vanguard schema of enlightened rule' that reaches into despotism. He finds it a utopian construct and unlikely to attract new followers defecting from conventional urban lives. It is difficult and impractical to establish an alternative society within contemporary society, he argues, without a vision for accommodating the legal and economic forces of the wider world. Eco-communities can founder by being simply undesirable for creating 'what today's suburbanites fear would be lifestyle grounded on excessive immobility, labor, pain, powerlessness, and inconvenience' (p. 172). Luke also raises the issue that in creating an isolated utopian alternative city, there is little that can be applied to existing cities where the major environmental problems lie.

Three aspects of the successes and failures of eco-communitarian projects stand out from a practical perspective, all of which need to be satisfactorily addressed to ensure success (whether rural or urban). First, the internal dynamics are critical, so that a system of governance needs to be in place and the best of these ecovillages exhibit both effective leadership (which can be collective) and some democratic or participatory mechanisms. Second, there are established ways of dealing with the outside world in ways that do not degrade the eco-community. Scholarship on eco-community theory has tended to emphasise the role separateness and independence, but in practice, it is the connections with conventional practices (such as economic, employment and transport services) that can be as important for success as the disconnections. Third, although it is often neglected in scholarship, success in eco-communities appears to also depend on having a successful internal economy. Ecovillages need to have a sufficient generation of resources, in terms of produce and monetary income, to meet collective and individual needs for goods and services. This is not to say that eco-community economies resemble conventional monetised communities and households in capitalist economies, but eco-communities do have economies and these cover income, exchange, production, consumption and redistribution.

Eco-Communalism as a Solution to Urban Transport Problems

Community-owned transport has a connection with eco-communalism across these central ideas and can represent its ideals in action. Much of eco-communalism, however, concerns utopian visions for new societies that entail striking out anew or fitting a bioregional consciousness over existing cities and their hinterlands. As was noted by Luke (1997) above, in the case of the former, such efforts leave the cities where these problems are rooted to sort out their own solutions and the latter, bioregionalism, seems utterly at odds with the complexities of the actual ecological footprint and administrative systems of large cities. Bioregionalism has a distinctly rural character, so that its relevance to the problem of organising urban transport requires interpreting eco-communalism in a distinctly urban context. Necessarily, some key features of the closer association between the natural and social spheres of bioregionalism must be downplayed.

There are several potential benefits of ecovillages in addressing transport's environmental and social costs, many of which arise from the relatively small scale of settlement, closeness of community relations, provision of local services and concerted efforts to minimise material consumption. It should be noted that there has been very little empirical research into the mobility aspects of ecovillages and most insights into this aspect of ecovillage living come from broader and more qualitative descriptions of general ecovillage functions and design. These key ecovillage characteristics create conditions that tend to reduce total travel demands, reduce travel distances (at least within the settlements) and create opportunities for greater use of walking and cycling compared to conventional car-based mobility in developed nations. Travel distance is reduced where ecovillages cluster travel destinations within smaller settlements, such as for employment and production centres, commerce and exchange, education and meeting places. Transport is a major expense for households in industrial societies, so obviating the use of private motor vehicles saves mobility expenditures for ecovillage households.

Street design in rural ecovillages is often designed around local needs, rather than for the primary use by motor vehicles; in urban ecovillages established within existing city landscapes, efforts are made to use streets for a variety of purposes and to restrain motor vehicle use. Much of the movement for shared space using urban streets, as widely practiced in Northern European cities, draws on the ideas of eco-communities in this regard. An obvious advantage of low motor vehicle speeds within settlements is that roads and streets can be smaller and less disrupting of other activities and pose lesser threats to other road uses. With closer social ties usually being an objective of eco-communities, there is greater interest in addressing mobility challenges through local collective solutions. Opportunities for sharing motor vehicle use, for example, are likely to be higher in ecovillages. Special accommodations can also be made for walking and cycling in ecovillages, including bikesharing schemes.

Over and above these aforementioned circumstances and ecovillage design features favouring more environmentally sound mobility choices, is the ethos supporting such choices and actively rejecting (or at least, not encouraging) private motor vehicle use. Such communities will not adopt motor vehicles as a primary solution to mobility needs because the conventional symbolic values associated with ownership and use, such as appeals to social status, self-affirmation or the excitement of speed and high performance.

However, the strengths of the independence of eco-communalism effectively ends at the points at which local services mesh with existing public transport services, whether as operated privately or by the state. Maximising the efficacy and environmental and social benefits of community-owned transport requires it to have some aspects of integration with larger transport systems. Longer and more complicated journeys should be taken by the most environmentally sound modes and in nearly all cases that will entail using mass transport modes. In order to minimise the ecological footprint of urban mobility, it will be necessary to possibly compromise the values of some approaches to eco-communalism by enabling integration between community-owned transport and public transport.

From the perspective of sustainable transport, eco-communalism has not presented any particularly notable advances to the usual dialogue, other than demonstrating that transport with low environmental impacts can be achieved under the right circumstances – this may be partially the outcome of insufficient investigations into this aspect of ecovillages. In a good number of eco-communities in developed nations, the choices and use of transport does not appear to necessarily differ from conventional suburban living. Transport practices in some rural and outer-suburban ecovillages could even exceed the environmental impacts of conventional urban dwellers using only public and active transport in large cities. There are, however, a number of dimensions of eco-communalism in theory and practice that could serve the project of community-owned transport.

First, there is the eco-community contribution to alternative finance models. There are, of course, a great many aspects to the emerging 'green economy', but here we are interested in what can be termed grassroots sustainability enterprises. Such exchanges and contributions form an essential role in building the social bonds within these communities. Eco-communities are founded on the voluntary contributions and exchanges of time, resources and skills of their communities outside of the market economy; how these are arranged and managed is of particular interest in the era of an emerging 'green economy'. Three aspects are of particular interest, that of community banking, bartering and alternative (or complementary) currencies, the latter of which has attracted wider interest with the creation of bitcoin, an online peer-to-peer payment system. Community banking is widely practised around the world and in its simplest form is a finance institution that is owned and operated locally. In the context of eco-communalism, community banks are used to provide financial support for both community projects and for individual households. Although ecovillages may seek to distance themselves from commercial life (but noting that many are actively engaged in a range of commercial activities, such as farming and education), they often need to make major investments in capital equipment, particularly for installing larger-scale renewable energy systems and for water storage and treatment. Many models for community banking are in use, but most have a basis in share ownership by members of the ecovillage either individually, or as a collective.

Rather than using the prevailing national currency, some ecovillages have developed alternative systems to replace the use of money. Seyfang (2009), who has studied the field of complementary currency extensively, describes an array of such mechanisms, including local exchange trading systems, green saving schemes and time banks. Community members can conduct the exchange of skills and services, acquiring debits and credits through the trading system. Time banks offer a way of exchanging labour, but differ from free market economies because in time banks all labour tends to be valued equally regardless of the tasks involved. Participants can exchange labour and assistance using a 'time broker' to produce what Seyfang (2005) calls a 'reciprocal volunteering scheme'. He notes that it is particularly used by those outside or on the margins of the wider economy, such as the unemployed, for tasks such as small do-it-yourself jobs, gardening and providing lifts, and time banks aim to engage those who are otherwise excluded. Benefits flow in many directions; recipients receive valued services without having to spend their own money, providers gain 'self-esteem and confidence' and there is a net gain to social capital as community ties are broadened and strengthened (Seyfang 2005). Complementary currencies are based on the notion that the medium of exchange, whether monetary or some other form, exerts a social influence on that exchange. These currencies are only valid within a specific community and are not recognised outside the community. Such alternative currencies operate in conjunction with the monetary system, i.e. they are complementary, but enable other sorts of exchanges not possible using money. Well-known examples are Scotland's Findhorn Foundation and their currency, the Eko and Italy's Damanhur that uses a currency they call the 'Credito'.

Secondly, there is the social dimension of ecovillages that provides lessons for the community provision of transport. Ecovillages are built on creating and strengthening the bonds between people with a common interest. Most ecovillages have decision-making systems and procedures built deliberately along democratic principles as determined by each community in accordance with its circumstances. To survive and flourish, ecovillages need to make good decisions for the interests of the community, to evaluate past decisions and to take corrective measures when needed. There is a tolerance of diversity of views and recognition of individual needs, such as for children, working parents and the elderly. Critically, the sharing of resources and assets of the community is central to the ecovillage life, as is the need to share labour and specialist skills. Many ecovillages consciously promote education and learning, for some ecovillages, programmes for the wider community provide sources of income.

Thirdly, ecovillages have accumulated experiences in ecological design and construction. Ecovillages are typically of a scale and design so as to aim to achieve a degree of independence and self-sufficiency in basic services and resource needs, notably residential energy supply, water supply and wastewater treatment and food supply. Often these service systems are communal and scaled to meet the needs of the entire community. Materials for construction are often locally sourced and from renewable resources or abundant non-renewable resources. Manufacturing and fabrication may be done in local workshops. In addition to private residences, ecovillages typically also have communal buildings and services (such as shops), which in larger villages may include education, recreation and medical facilities.

Taken as whole, these features can provide conditions suited to collective approaches to local transport. In short, ecovillages have experience in financing collective enterprises and have ways of paying for such services beyond that of monetary economics, including trading systems and complementary currencies that could be applied to local transport services. Collective ownership, shared resources and assets, institutions for community design-making and communal management are features of ecovillages that provide the sort of foundation institutions on which community-owned transport can be built and subsequently operated. Experience in eco-design might also lend itself to community-owned transport. Ecovillages have been test-beds for a variety of technologies and experimentation in their use; further, they have an interest in integrating different services and systems. Integration of transport services into other services and productive activities could bring many benefits, such as offering a broader array of financing choices, of providing necessary labour and for using transport to promote other productive activities.

6 Urban Transport as a Commons

An important foundation of community ownership is understanding the resource that is communally owned and the proposition that urban mobility, or at least forms of urban transport, constitutes a resource that can be owned communally. Certain characteristics of mobility make it a *commons* resource and amenable to community ownership under suitable management regimes. By knowing something of urban transport as a commons, particular requirements of community ownership can be revealed, which in turn, provides insights into the types of community ownership approaches that may succeed.

Common Pool Resources and Common Pool Property

Common pool resources ('commons') are those that cannot be owned but can be accessed by a community or can be collectively owned and managed. Common pool property can be owned privately or collectively and may be accessible to others. Seen through the lens of property rights (i.e. enforceable rules of resource access and use), commons denote collective resource ownership or the rights of access and use of certain resources owned by others. Property rights are a bundle of specific rights and cover such matters as having exclusive resource use, resource services use, transforming the harvested resource/service, excluding other users and transferring/selling of rights of use and applying of harvesting technology (Ostrom 1990, Buck 1998).

Economists have developed their own definition of common pool resources using two conditions. Under the first condition, commons are resources subject to rivalry between actual or potential users, meaning that consumption (i.e. use) by one user deprives another's. Under the second condition, it is difficult or impractical to control access to the resource. Commons, therefore, are depleted by consumption and are available to be consumed unless some measures are taken to control access. Confusion sometimes arises between commons and public goods, especially in general usage, a matter made worse when it is assumed that the concepts are interchangeable. This may have arisen because of the phrase describing desirable social benefits as being 'in the public good', but that is not what is meant here, where a public good might be equally likely to be harmful as beneficial. In the world of scholarship, these are different types of resources and it is helpful to be clear about this. While commons and public goods are both essentially available to any potential user, the use of public goods does not diminish the resource, such as enjoying national security or admiring a rising moon. However, sometimes this distinction is hard to draw and in some instances, the same resource may be classified one way or the other depending on circumstances. In practice, disputes over shared resources nearly always concern commons, not public goods.

Commons form a wide variety of resources. Examples of natural resource commons include tangible things such as forests, grazing fields, wild animal herds, fisheries and water bodies, as well as service-providing resources, such as genetic material and the climate system. Most global commons are open access resources and vulnerable to over-exploitation, such as the open ocean, global climate system and global biodiversity (Buck 1998). It is also worth noting that sometimes it is not the direct benefits from the commons that are sought, but an indirect benefit, such as agricultural production made possible by a shared irrigation system.

Commons may or may not be in the form of common resource property. Natural resource systems and processes are not likely to be owned, nor are ecosystems (genetic materials are another matter, however). Government ownership is extended over many commons, such as public land and the outputs of government research. Knowledge itself can also be a commons when shared socially, including historical knowledge and folk wisdom; unless it is held as private property, science and its outputs form part of this knowledge commons. A portion of this body of knowledge is used for managing resources and this is a piece of the wider knowledge and practices that constitutes culture and tradition.

In effect, many natural resources that are now private or state property are appropriated common pool resources, such as occurred on a large scale during the European colonisation of the new world and the enclosure movement in eighteenth century England. Examples of social resources are legal systems, parks, reserves and irrigation systems. As Bollier (2013) states in a work that seeks to defend commons against appropriation, commons include natural resources, public knowledge, academic knowledge, culture and open spaces, the Internet, government information and drug research.

Technological and other changes provide new incentives and opportunities to extract commercial value from commons, such as from genetic material, prompting public debates over efforts to create private property rights over such commons resources. Similarly, environmental problems from resource depletion and pollution (which, from a commons perspective are the same problem) bring forward new efforts to impose management regimes over resource systems that are often without controls over their use or appropriation. Although the commons concept is not always evoked in disputes and debates over these issues, the essential political conflicts are usually based around contesting claims for use and management of commons, such as establishing traditional rights of access, the need for adequate commons protection, the legitimacy of claims for use and the accountability of resource managers to a wider community with an interest in the commons resource.

Institutions for Commons Management

In indigenous, traditional and pre-modern societies, common pool resources have furnished the basic materials for social survival and flourishing for the entirety of human existence in so far as we can tell. When records are available, commons management regimes feature strongly in the longest-lasting social institutions. A great part of social organisation, covering language, culture, religion, taboos and relationships with other groups, is linked to forming and maintaining institutions for managing commons resources. A primary objective of all social organisations is solving collective problems and managing the common pool resources on which communities and households depend. Failures in these institutions, for whatever reasons, can result in the collapse of societies (Ostrom 1990). Using, protecting and maintaining commons occurs through common property regimes that entail a wide range of institutions, collective practices and social norms. Commons property regime management aims to protect the resource by regulating resources in order to provide the flow's on-going benefits without degrading the stock of the common resource. Further, a single commons might be managed by several institutions and by a mixture of private and communal owners.

Commons without any controls over their use are 'open access resources' and vulnerable to over-exploitation. Open access resources are without the protections of a common property regime and can be harvested by all those who choose to do so; there are no property rights, no controlling authority and the outputs are generally harvested for sale, rather than use by members of the commons. Failure to recognise the critical distinction between open access and common property regimes lays at the foot of the popular misconception that all commons are vulnerable to overuse in the so-called 'tragedy of the commons' as expressed by Garrett Hardin (1968) (namely, that scarce resources open to all will be exploited to the point of exhaustion, thereby bringing ruin to all users). Hardin's argument went on to become one of the key rationalisations in debates over natural resource use.

Although the 'inevitable tragedy' argument is now discredited, two other related misconceptions continue to plague commons thinking, argues Bollier (2013) and certainly these concepts appear frequently in the commons literature. First, there is the prisoner's dilemma based on game theory that is used to prove the unlikely prospect of cooperation where there are incentives for personal gain and it 'purports to show that cooperation is usually irrational and unlikely to solve collective action problems' (Bollier 2013, p. 19). Second, there is Mancur Olson's logic of collective action thesis that argues that self-interested people acting rationally will not act collectively because of the preference for free riding if the user cannot be excluded and there is no incentive for contributing to maintaining the commons. Both of these ideas reflect the faith that conventional economics has invested in *homo economicus*, with its core assertion that human behaviour is understood as that of self-interested amoral individuals seeking to maximise individual economic benefits independent of any ethical beliefs and existing outside of any social context.

Of course, there is more to life than economics and the dismal science offers a distorted and narrow view of social life and individual behaviour, as clearly demonstrated by the range and breadth of individual cooperative and altruistic behaviour towards collective projects. Significantly, Hardin's argument was conceptual speculation that was easily refuted by the wealth of empirical evidence, as collected by Elinor Ostrom and others, demonstrating that commons essential to local communities are carefully managed through various social institutions and practices. It is a small semantic change, but one of great importance: what Hardin speculated about was the tragedy of open access commons. Ostrom notwithstanding, these problems are symptomatic of the limits of conventional economics in understanding natural resources and its neglect or incomprehension of social and ecological phenomena that provide natural resources and the social institutions, cultures and human behaviours developed for resource harvesting, managing and maintenance.

In place of a tragedy, Rose (1986) found 'the comedy of the commons' and refuted the spectre of the commons' tragedy by arguing on legal grounds (using British common law) that some resources are 'inherently' communal, such as roads and waterways. Rather than being degraded by additional users, such commons can be distinguished by providing additional benefits from increased numbers of users, using the example of the market square, where more users (both vendors and customers) lowers the costs of use. Having roads open to users, Rose argues, was essential to commerce being maintained and increased and, as Frischmann (2012, p. 7) points out, 'commerce is itself a *productive activity* that generates significant positive externalities'. Frischmann (2005) argues that because free markets can fail to satisfy demand for some resources/services from infrastructure, commons are necessary to avoid such market failure. Network goods exemplify this effect, wherein simultaneous consumption increases the benefits of all users through the benefits of scale, such as a telephone system.

Three broad responses are available to the problem of common pool resources: government ownership, private ownership and communal ownership (Ostrom 1990). Ostrom devoted considerable attention to the design and functioning of these resource management schemes. Efforts to derive a consistent set of explanations for the relationship between the type of common pool resources and associated management regime have been frustrated by the sheer variety of historical experience across cultures and circumstances. All sorts of things that we might not consider to be common property have been treated as such, such as children and personal belongings, so that the field has been resistant to being fixed by determining causes.

Case studies have been the mainstay of commons research, with greatest attention given to natural resources management in indigenous and traditional societies and to natural resource harvesting regimes, such as for irrigation water, fish and wildlife. As such, the arrangements for regime management tend to be culturally based in specific communities dealing with specific local resources. Determining the factors for successful commons management is a difficult task but notwithstanding these difficulties, Ostrom (1990) in her *Governing the Commons*, developed eight design principles; Ostrom's measure for commons protection is whether there has been a continuous flow of benefits from the commons. These principles are as follows:

- Clear boundary definition of both users and resource system.
- Rules used for governing the use of collective goods are matched to local circumstances.
- Rules are designed, at least partly, by the local appropriators.
- Monitoring (by individuals) accountable to local appropriators.
- Graduated sanctions applied for appropriators who break the rules.
- Conflict-resolution mechanisms that are quick and cheap.
- Minimal recognition of rights to organise.
- There are nested enterprises, with large organisational units built from smaller units.

Subsequent scholarship has produced a critique of these principles and several modifications have been suggested. Cox *et al.* (2010) reviewed such studies and found them to be empirically supported. They comment: 'the primary role of the design principles is to explain under what conditions trust and reciprocity can be build and maintained to sustain collective action in the face of social dilemmas faced by CPRs [common pool resources].' A number of criticisms were that in practice there is greater fluidity in arrangements, such as in setting resource boundaries and the malleability in forging and enforcing rules. Three specific critiques were identified. First, that there are important aspects missing from Ostrom's principles that amount to an undervaluing of important social mechanisms, such as legitimacy and transparency. Furthermore, resources are not only defined by local factors, but can also include outside factors. Second, there is scepticism over whether these principles can be scaled up to deal with global issues and more complex networks. Thirdly, there are concerns whether the principles are generalisations that do not hold for all cases.

Although Ostrom's aforementioned research dealt with the use of commons, her later work began to consider the role of commons in production and more recent scholars have turned their attention increasingly to this aspect of commons. Frischmann (2005) identifies commercial, public and social benefits from commons and uses this to guide decisions for optimal management. He finds that where commons are valued for public and social ends, then communal management regimes are the preferable approach to management (because determining market values for the benefits in question is difficult).

Through history, the general trend has been the loss of commons under communal ownership to state and private ownership (Buck 1998, Bollier 2013, Wall 2014). This process of appropriation has been rationalised in a variety of ways, such as Hardin's aforementioned proposition that only privatisation can service competent protection. Other conventional economic rationalisations draw on the imperative that resources be used to maximal economic potential which requires bringing them into capitalistic markets through private ownership.

Needless to say, private ownership can result in the partial or complete loss of common pool resources, especially as markets favour short-term rewards and the economic reckonings rarely include the externalities of industrial production. For traditional owners or custodians of commons that are lost to private or state gain, the effect can be the complete loss of livelihoods, cultural identities and even survival (Bollier 2013).

When considering resource use and commons, there is a frequent tendency to adopt an all-encompassing approach to the natural and social worlds. Viewing the world through an economic lens, only resources under property rights of some sort (either communal, private, state or co-joined) and without property rights (*res nullis*, i.e. unowned) can be seen. In effect, such an accounting brings everything into the frame of economic markets. Such a reckoning is an instrumental valuing at its greatest; there is no place for Nature having an intrinsic value or indeed, for social resources to also be valued in their own right. Untrammelled anthrocentrism, namely only recognising on instrumental values, is basically inconsistent with an expansive approach to environmental justice. However, it should be possible to pursue the cause of communal ownership for commons without having to reject recognising that commons can also have intrinsic values or that there are phenomena that need not even be valued as potential resources.

Infrastructure Commons

Commons scholarship has been expanded to apply to emerging problems and to new technologies, particularly relating to digital media and the new communication platforms, but also to such phenomena as the use of renewable energy systems to maximise social and environmental benefits (Byrne *et al.* 2009). Although, as Frischmann has stated (2012, p. 3): 'Infrastructure and commons are not typically thought to be related to each other.' Following the basic definition offered above, these infrastructure resources (and/or infrastructure services) benefit communities, are diminished by additional users and excluding those additional users is difficult, so these resources/services qualify as commons. Infrastructure resources in modern industrial states are available to the community without discrimination (i.e. use does not depend on user identity or intended use) (Frischmann 2012). Infrastructure commons is a growing field of study, but it remains considerably smaller than the body of work (much of it based around anthropology) on traditional and indigenous resource harvesting and producing systems. Ostrom's aforementioned commons' 'rules', for example, were mainly drawn from such traditional commons case studies, but these more recent studies of global problems, industrial infrastructure and contemporary technologies are bringing forward new sorts of problems and challenges to established thinking about commons.

Contemporary interest in the commons is being pushed forward by two particular problems, those arising from global environmental problems and those from large infrastructure investments, particularly in the realm of urban services, the latter of which is of particular interest here. Large urban infrastructure systems function as commons and provide services essential for maintaining urban life, such as transport, water, wastewater and energy. Social and technological changes are creating new phenomena that provide resources for consumption and where there can be difficulties in managing that consumption so as to preserve the service, such as the Internet (e.g. Hess and Ostrom, 2003). Generalisations applied to traditional infrastructure (such as the necessity of government involvement, open access and generation of wider benefits) are being challenged by these new developments. At the scale of system components, there is a mixture of all sorts of property and resource ownership and management arrangements, but this does not alter the commons character of these systems at the broad scale.

Challenges to established views on commons include the difficulty of establishing the geographical boundaries of large-scale commons and determining the conceptual boundaries of large environmental commons and technological systems. Phenomena such as biodiversity, the Internet and the ozone layer challenge even the usual language of definitions typically used to legally define property rights. Similarly, the close tie between commons and community organisations in traditional societies does not hold for this new group of contemporary commons issues that involve dispersed, diffuse and indeed, virtual resources and services. Furthermore, both the resource users and management institutions have similar characteristics as these resources, such as being national, regional or even globally dispersed. Identifying the success factors behind this new breed of commons issues is more complicated and differs from the local natural and cultural resources used by local communities as described and analysed by Ostrom and others.

There are four essential functions for maintaining infrastructure commons in circumstances where overuse can occur, according to Künneke and Finger (2009). These authors draw on European experience and circumstances where many public infrastructure systems are being broken up through privatisation and corporatisation. First, there is system management that entails coordinating the infrastructure system in the short term. Economic trends can mediate against coordination with decisions (Künneke and Finger 2009, p. 9): 'With growing fragmentation of the technical systems because of unbundling, outsourcing, and the like, there is therefore a growing need to coordinate all operations and actors involved.' Second, there is capacity management that refers to a conscious and strategic allocation of resources between users and their needs, noting that the authors identify strategic, tactical and operational allocations. Third, interconnection within the network is essential. Without the necessary physical, information and other connections between the parts of the network efficiency, reliability and the wider network benefits are foregone. Fourth, there is the need to operate networked systems effectively by ensuring interoperability between the components as this 'ensures that the elements of the network are combinable. In other words, interoperability defines the technical and institutional conditions under which infrastructure networks can be utilized' (Künneke and Finger 2009, p. 12). Exemplars employed by the authors include rail lines suitable for the rolling stock and air navigation systems providing effective guidance. Additionally, interoperability is integral to management as technical standards, regulatory controls and other institutional tools sets the conditions for resource users (covering market entry and exit).

Urban Mobility Commons

Included in this approach have been studies considering infrastructure, including transport systems, as commons (e.g. Frischmann 2005, 2012, Künneke and Finger 2009). Wills-Johnson (2010) used the commons concept as offering lessons for the collective use of Australia's railway infrastructure and Glover (2011, 2012) classified public transport as a commons. In essence, the stock of the commons is the transport infrastructure and associated services and the services derived are personal mobility. A commons for urban mobility conceptually covers the infrastructure, rolling stock and vehicles and related services, but this is complicated by the mix of private, state and communal stakeholders, the mix of resources under property rights (private, state and communal) and those of open access, together with the mix of benefits and services sought from mobility (namely the direct and derived benefits).

Public Transport Commons

Despite the familiarity of public transport as a concept and a transport system, there are surprisingly few comprehensive definitions of public transport and no single and widely accepted definition within transport scholarship. A review of definitions used reveals definitions drawing on ownership, institutions, modes, law and regulation (Glover 2013). Typically, public transport is described as a particular set of transport modes capable of providing mass transport, as transport that is not privately owned or that it is transport that is state owned. Clearly, these approaches tell us little about public transport and offer little guidance in differentiating between public and private transport, and corporatisation, privatisation and other such issues confound this uncertainty.

By using a commons framework, the identity of public transport can be clarified. Institutional factors to define public transport can serve to initiate this discussion; public transport has governance through public policy mechanisms, has financial structures based in government and has a primary objective of providing mobility. Two factors denote public transport as a commons. First, it is a consumptive service in which there is a limited capacity, so that potential travellers compete for that service. Capacity limits exist at the vehicular level and at operational levels, with the limits reached at the point of crowding. Second, public transport is also effectively unrestricted; services are not rationed and it is difficult to restrict access to the services. While pricing and fares influence the equity aspects of access, this does not alter the fact that nearly all public transport systems have public service obligations to be available to all potential users and, in democratic nations at least, the general public considers that it a right to access public transport systems.

Public transport networks are infrastructures providing a mobility commons and this infrastructure also occupies considerable open space within cities, itself a commons resource. Through history, new transport and associated technologies create new services and displace existing services. These developments have implications for urban mobility commons. For example, governments made allowances for building fixed rail infrastructure through existing city streets and thoroughfares to provide for public transport services, often to private companies in the first instances. Effectively, this was a privatisation of public open space, transferring open access into private hands.

Urban Roads Commons

Roads are an interesting problem in the context of mobility as a commons. Private vehicles compete for road space with other private vehicles and other classes of road user, namely public transport vehicles and active transport (i.e. cyclists and pedestrians). In most cities, private vehicles have won this competition and occupy the bulk of road space, courtesy of public policy decisions that formally designate the use of road space between vehicle types. When there are no restrictions on use of this open access resource, motor vehicles dominate in part because drivers and passengers are able to enjoy a number of perceived benefits (such as high travel speeds, vehicle comfort, convenience of use and status acquisition) and generally avoid the costs that their choice imposes on others, future generations and the environment (through pollution, road trauma and the like). Because motor vehicles increase the risks and severity of road trauma, in the competition for road space use, the vulnerable active transport modes are invariably discouraged as overall motor vehicle traffic increases. Road trauma rates internationally match the extent to which the vulnerable road users are mixed with motor vehicle traffic.

Here we find many solutions being applied. Generally, pedestrians have been provided with footpaths (sidewalks) within road reserves and increasingly the same is being done for cyclists in the form of bike paths (off-road) and bike lanes (on-road). In one sense, the communal road space is exclusively partitioned for private use by vehicle/mode type, rendering invisible the competition between the modes under the rationale of improving road safety (or decreasing road trauma, depending on your perspective). Separating the modes on roads does not necessarily deprive mobility, as there are efficiency gains for motorised vehicles by removing slower, non-motorised modes, but such decisions do privilege private motor vehicle users over competing users. Freeways present a different sort of example; whereas urban roads and streets might have originated for use for many types of vehicles and serve as open spaces, freeways are purpose-built for motor vehicles. By design, freeways carry high volumes of traffic and act as a public subsidy to mobility using motor vehicles. Subsidy arises because governments allow the transfer of various commons resources from the wider community to the community of car drivers, such as the urban space used for the infrastructure, air and waterways that can be polluted and absence of noise. This transfer without compensation is from the community who enjoyed the benefits prior to the freeway's construction and operation to the motorists and the infrastructure owners, who might be either government or corporate.

There have been demand management solutions as well, such as road tolls, road pricing and congestion charging, which in one sense are rationing measures made possible by creating property rights over roadways in which governments exercise the rights of ownership. Privatisation, for example, provides a limited solution to these problems. Restricting use through privatisation, with the obvious outcome that allocating use rights through pricing mechanisms, is financially regressive (meaning that it disproportionally disadvantages those with fewer means to pay). It is not difficult to restrict car use by causing driving to be more expensive and making road use into a commercial service overseen by the state, but those at the lower end of the socioeconomic spectrum will proportionally bear the greater burden (based on their weaker ability to pay).

One social movement of particular interest here is the Dutch woonerf ('living yard') street concept that has now spread across Europe (including Russia), with similar versions in the UK (known as home zones) and elsewhere. In The Netherlands, woonerf are streets in residential areas featuring very low speed limits (15 kph) for vehicles and with other traffic calming design measures designed to limit car use and encourage use of the street as open space catering to multiple uses (i.e. there are no separate footpaths) (Carr *et al.* 1992, Beatley 1999). Introduced in the 1960s, woonerf streets are a deliberate effort to provide a safe setting for children's play, walking and cycling. Car traffic is limited and car parking is meant to be unobtrusive. There is legislative support for woonerf measures (passed in 1976 and updated in 1988). These streets include tree plantings (including in places where this was not possible when car access was prioritised), space for benches outside homes and full street access for pedestrians, with street lighting designed for social needs, rather than for motor vehicles. Woonerf are restored commons that encourage a range of open space uses and reap a range of benefits that cannot be achieved under conditions that favour road traffic, such as spending time outdoors, socialising between neighbours, children's play and adult recreation, as well as promoting walking and cycling for adults and children. By trading off the optimal conditions for motorised transport to gain access to open space, a narrow allocation of benefits is replaced by a wider range of benefits enjoyed by a more diverse representation of the local community. From this exemplary practice, the presumption that the commons of residential streets are best allocated to private gains by motorists is revealed as unpopular and inimical to a number of social goals, such as having safe streets, encouraging physical activity, promoting independent children's mobility and facilitating active transport.

Part III

Community-Owned Transport Practices

7 Community-Owned Transport

Overall Concept

In drawing together the concepts and key ideas explored in the preceding chapters, this chapter describes the concept of community-owned transport and explores the case for community ownership, including the ways in which it might promote sustainable urban transport. In doing so, the chapter also argues the case of community ownership based on a critique of the dominant approaches of private and public transport in delivering urban mobility. As stated above, a very broad interpretation of the concept is offered: community-owned transport is a mobility service provided or facilitated through a community-based organisation located within civil society. Community ownership can take a variety of forms, including cooperatives, unincorporated and incorporated associations and companies of various types. No restrictions are considered here in terms of form of community organisation, membership or structure of the organisation, whether stand-alone or within another organisation, values or ideology of the organisation, relationship with other organisations or type of mobility service offered.

A Basic Case for Community-Owned Transport

Moving people within and between the cities of the developed world, and increasingly in the developing world, in ways with low social and environmental costs necessitates shifting them out of private motorised vehicles and increasingly into shared vehicles and onto their feet and bicycles. In much of the developed nations, the rise of the motorcar meant the demise of public transport and walking and cycling. By way of contrast, in a number of Western European nations, public transport continued to play a major role in urban passenger transport, but even here, the rate of motorcar use increased after the war. As we know, the post-WWII era has been the age of the motorcar, a period of automobility. But now that era might be drawing to a close, although in ways and for reasons that were not envisioned until quite recently with the rise of the need for sustainable transport.

Urban transport has been largely a function of transport technologies and their associated technological systems. Social, political, economic and environmental costs and benefits of urban transport have their beginnings in the technology choices made in pursuit of urban mobility. Different cultures in different locations make their mobility choices according to their values, preferences, institutions and other factors, but their options are a function of the available transport technology choices. When new forms of transport technology are invented and become available, new ways of moving people around cities become possible. Looking at the history of urban transport, we see an interesting progression.

In many ways, the motorcar has been a victim of its own success. Yet, the end of automobility is in many ways a development at odds with the general sweep of transport history, wherein we expect to see changes in transport technologies as the driver of changes in transport systems. So the revival of public transport is of interest because not only is it not a change in direction that is not being driven by technological change, but it also marks a return to the modes that were deemed redundant by automobility. Mass public transport's revival is not due to technological change, but one driven by the politics of sustainable development and the challenge of sustainable urban transport. Realising public transport's return as the dominant means of urban transport worldwide is a transition that necessitates actions in the broader social realm, involving politics, institutions, governments, corporations and the broader community. Urban transport systems necessarily involve governments at every level, but in the past government decisions were coupled in various ways with markets and economic signals.

Yet we have relatively few insights into the politics of transport policy and practices. Political inquiry has delved into many of the realms in which transport operates, notably urban studies and in historical accounts of cities, but transport itself has been largely neglected. (Several works on transport and public policy in recent decades bear titles referring to politics, but their content is confined to the machinations of public policy analysis and institutional matters.) There has been a more vigorous effort made in cultural inquiries into transport and the concept of mobility, particularly into cars and the idea of automobility, than there has been in politics, per se (e.g. Sachs 1984, Ross 1996, Seiler 2008, Featherstone *et al.* 2005, Sheller and Urry 2006).Most authors on transport, including those examining public policy, either make a deliberate effort not to raise political issues or do so largely unwittingly, confident that comprehensive-rationalistic views express a universal desirability. Exceptions to this generalisation are usually concerned with the politics of motorcars and road building, such as Paterson's (2007) *Automobile Politics*.

With the rise of sustainable transport this situation has changed somewhat, but the resulting scholarship has continued to have an emphasis on institutions and public policy rather than on politics (e.g. Dudley and Richardson 2004). Many of the large infrastructure systems essential to modern industrial life have been opened to political studies, such as energy, electricity and water resources, but transport has tended not to have been subject to this sort of scrutiny. Transport operations and management are dominated by tools, concepts and insights from engineering and economics and much of the scholarly writing on urban transport has either knowingly or subconsciously framed its inquires within these paradigms of knowledge. Studies from these disciplines examine issues that are typically downstream from the political factors and circumstances that gave rise to these issues and refrain from engaging in such subjective matters. Decisions and practices over transport technology choices, transport energy, the distribution of services, access and equity in service provision, the ownership of transport and managing and controlling of transport operations are clearly not engineering and economics topics. In democratic and open societies, such essential questions of these decisions and practices need to be examined in the light of politics. Rather than continue this general approach to urban transport that eschews political values and questions as being unnecessary and irrelevant or divisive and controversial, this work considers that social science studies into urban transport need to engage with the politics of transport.

Despite the consensus about the general desirability of the idea of sustainable transport amongst the governments of the developed nations and their cities, the ready transition from automobility to sustainable transport is far from being either assured or indeed, immanent. This uncertainty over the revival of public transport comes at a time when several major social trends intersect, most notably the aforementioned widespread adoption of neoliberalism across the developed nations as the prevailing ethos of the dominant political practices and processes of public policy formation. Although sustainable transport requires public transport to become the dominant means of passenger travel in cities, this requires governments to promote the growth of public transport services. Such government support entails changes in public funding. Certainly, the obvious change is to greatly diminish the public funding of roads and freeways, to require motor vehicle users to meet more of the costs of motorcar use and to end other public subsidies of the motorcar transport systems and to direct greatly increased financial and institutional support towards the goal of greatly increasing public transport services and usage. Extensive public spending on public transport sits at odds with neoliberal economic programmes, placing governments in a contradictory situation.

There are also other political impediments to such a course of action. Put simply, the stakeholders in the motorcar systems are of considerable political importance, representing as they do the major industries in the petroleum, motor vehicle manufacture, motor vehicle retailing and serving, motor vehicle insurance, road and freeway construction, motoring organisations and others (Paterson 2007, Urry 2007). These industries have developed close and intricate relationships with governments and such ties are not easily undone. Even when governments escape such influences, there are many other sectors and projects competing for public sector support in health, education and social welfare. In societies with high levels of car ownership, there can be a majority of the electorate enjoying the public subsidy of motorcar use and who are willing to use the ballot box to express their displeasure at elected officials seeking to oppose any lessening of motorist's enjoyment of public largesse. As a result, in representative democracies, switching public funding across to public transport is far from easy and around the world, many public transport systems have not been able to gain access to the funds needed to expand their services.

So we can find an obvious set of political barriers for governments to promote sustainable transport. There is also another set of barriers, namely those arising from the existing institutional arrangements governing public transport.

Essentially, nearly every urban and regional public transport system in the developed world is either owned and managed by a public authority or owned and managed by a corporation under government oversight, or some form of mixture of the two. Public transport is usually owned and managed as a monopoly; there are certainly advantages from monopoly suppliers, but there are also costs and losses. It cannot be axiomatically assumed that all public transport operators simply favour greater efforts in sustainable transport as significant public policy change always shifts the existing distribution of power, influence and resources within the mix of public entities.

An alternative institutional approach is a different form of public transport ownership and management, that where community groups own and manage their own transport services, i.e. community-owned transport. Here, we look at what community-owned transport means, how it works or might work and how it can contribute to sustainable transport.

Making a Case for Community Ownership

Urban Mobility as a Commons

Urban mobility and urban transport systems are commons. Historically and with considerable consistency around the world and over time, modern transport systems (i.e. meaning motorised mass transport following the Industrial Revolution) have featured centralised state control, notably over roads and public transport. Private property rights have been extended over motor vehicles and in recent times there has been privatisation of various components of public sector transport systems, such as toll roads and franchised public transport services. As commons, much transport infrastructure is open access, notably roads that are subject to congestion as a result of over-use of the resource. (Noting that congestion may be a more nuanced issue than usually thought.) Roads have been left generally as open access with state control only recently moving towards demand management strategies, including rationing, pay-for-use and the rise of private freeways. Active transport infrastructure for walking and cycling is also open access, but rarely considered in such terms.

Loss of these commons can occur in a variety of ways and oftentimes such losses can pass unnoticed until a specific complaint is made. David Bollier's (2013) *Silent Theft* identifies five types of loss from enclosure of the commons:

1 Public asset streams are lost and diverted from causes in the public interest.
2 Enclosure fosters a concentration of market power because it is mostly undertaken by the largest firms and produces higher consumer prices and less market competition.
3 Short-term exploitation is favoured over long-term stewardship.
4 Citizen rights are diminished, as is public accountability.
5 Market values are imposed in realms that ought to be free of commercialisation (such as community values, public institutions and democratic processes).

It follows that strategies for continued privatisation of transport commons will incur a range of costs and it seems likely that many would evoke considerable political and community opposition. Further, many of these costs seem directly in contradiction to the aspirations of sustainable transport, making further use of commons appropriation strategies self-defeating.

Dealing with the commons of urban mobility, the inadequacy of such market-based solutions leaves open the entry of other arrangements, such as community ownership; there is nothing in the inherent characteristics of the transport commons that excludes a place for community ownership. Within the existing transport systems there are unmet needs, indicating market failures, due to such factors as the limits of public transport services and the costs of private vehicle ownership. In theory, therefore, there is a 'space' for community ownership and there would appear to be mobility needs that can potentially only be provided through community ownership of mobility services.

Urban Mobility and Collaborative Consumption

Collaborative consumption is a new and growing phenomenon in mobility around the world and assumes a wide variety of forms and arrangements. These developments have created new options for urban mobility, such as the aforementioned carsharing, ridesharing and bikesharing schemes. Technologically, collaborative consumption has prompted innovation in mobility, such as the examples described in a preceding chapter. These innovations have created new opportunities for such services as integrating public and private transport journeys and providing for short-term car leasing. Both in a commercial and institutional sense, collaborative consumption is helping to break down the strict differences between public and private transport and giving rise to hybrid institutions providing mobility.

Some critics might regard collaborative consumption as a sideshow to the main event of private car ownership and public transport because of its minor status in overall mode share, but there are reasons to consider its future more openly. Collaborative consumption in transport has grown at a rate that has been largely unpredicted by researchers and commentators and continues without any signs of slowing at this time. Over just a few years, fundamental assumptions about urban mobility have been overturned. Even if some transport researchers remain sceptical, several of the major automotive manufacturers have taken the concept seriously enough to invest in their own collaborative mobility businesses, covering several different approaches. It would seem both premature to consider that collaborative consumption will remain as minor mode share worldwide in coming decades and to also think that the evolution of urban mobility services has reached the 'end of history'.

Institutionally, collaborative consumption is prompting changes in public policy, legislation and the functioning of transport agencies to accommodate changes in transport services. Most collaborative consumption is firmly within the private sector, although much of this activity seems conceptually quite close to community-based institutions and several of the marketing campaigns evoke notions of community (noting that some of the key collaborative consumption transport providers are not-for-profit enterprises). It seems but a small step in many instances to swap out private ownership for a form of community ownership. It also seems likely that there will be new forms of institutions developed in response to these industries as new business models are created and as technologies develop further.

Promoting Sustainable Urban Transport

Credibility for community-owned transport contributing to the goals of sustainable transport rests on two requirements that are potentially in opposition. On one side, there is the ecological justice goal of improving mobility access and equity, especially improving the lot for those with fewest resources and fewest opportunities for access to mobility services. On the other side is the environmental protection component of ecological justice that entails reducing the net and local environmental impacts of transport through such strategies as reducing motorised mobility, switching to low-emission vehicles, increasing public transport use and increasing active transport. For improving access and equity, the potential contribution of community-owned transport is straightforward, but the challenge is to fulfil this goal and simultaneously lower net environmental impacts.

Strategies and practices to ensure that community-owned transport assists in the transition to sustainable transport will be based on those aforementioned strategies and as canvassed in Chapter 3. There are several ways in which this could be achieved. Community-owned transport can fill the gaps in existing public transport services (covering both time and geographical coverage) and provide travellers with access to public transport and make public transport more used, more usable and more viable. Carsharing schemes can provide access to low-emission vehicles, ridesharing can lower the size of the motor vehicle fleet and bikesharing schemes increase active transport. If such measures are effective, then car use and dependency can be reduced.

Community-owned transport can also connect with many of the wider themes of urban sustainability and help in achieving its goals. Through its presence, community-owned transport organisations can bring forward key aspects of the sustainability agenda. These include such matters as the absence of safe walking and cycling routes in a community, challenging the assumptions that driving children to school is the best choice for children's wellbeing, promoting the use of locally-grown food and locally-available services, the need for local delivery services, and the requirement for local services in health, education, aged services, kindergartens and child care and local government. Creating such alternative mobility choices undermines prevailing assumptions in societies where automobility is taken as the model for urban mobility. As such, lifestyle choices and personal values linked to automobility are brought into a critical light, enabling some of the links between personal choices and environmental implications to be made more apparent.

Insights from Rural Mobility in Developed Nations

Research into rural mobility has been prompted by broader concerns with the aforementioned general uptake of access and equity inquiries in transport scholarship, coincident with the neoliberal public policies reforms and wider governmental concerns over public expenditures, highlighting the role of public subsidy in much rural public transport provision. Critically, the designation of 'rural' does not imply homogeneity, but includes accessible small towns, remote small towns and accessible and remote places. Rural mobility is relevant to community-owned transport in several ways; it deals with low volume, geographically dispersed, diverse and intermittent mobility needs across a broad social spectrum, although often in communities suffering socio-economic disadvantages and in locations where public transport facilities and services are usually sparse. Aspects of such mobility disadvantages can also occur within towns and cities, as previously described, but access and equity issues are often greater in rural areas. Rural populations have become the minority in developed nations and public investment in transport has favoured cities and towns throughout the modern era. For public transport providers, whether state or corporate, the central challenge of rural mobility is the obvious assembly of low and dispersed service demands (in space and time) that make cost recovery difficult and vehicle occupancies low. Rural public transport in developed nations is rarely commercially viable; public subsidy of services is the norm (with the fare box usually recovering between one-third to a small fraction of costs) (CIT 2008, Ellis and McCollom 2009, Laws *et al.* 2009). Interest in rural mobility has received far less attention than its urban counterpart, and although recent research has done much to redress this imbalance, it remains the lesser-understood issue.

In developed nations, the motorcar revolutionised rural life and it remains enormously influential; for many rural residents, normal life without car access is almost unimaginable and rural car ownership rates are higher than urban areas (DOT 2014, Mattson 2014). One factor that underscores the importance of motorcar use in North America, Western Europe and Australia is the general decline in local services in rural settlements, such as health, education, banking and retail services, necessitating trips to larger and more distant settlements. For rural residents with car access, trips of a few kilometres are easily traversed. For those without such access, destinations outside their own settlements reveal their mobility disadvantages. Woods (2005) notes that remote and isolated communities are more likely to have key services than similar-sized communities closer to larger settlements, but face longer trips to access those services not available locally. Ironically, a major factor in the historical loss of services in rural settlements was the advent of the motorcar. Increased rural mobility services and improved access, however, can reduce such economic disadvantages and improve rural economic prosperity (Farrington and Farrington 2005).

Mobility options for rural households – those without car access – are usually either some form of public transport, walking and cycling or a form of communal transport, a circumstance that creates highly inequitable mobility access. In the

US, 1 in 14 rural households do not have a car (about 7 per cent) but 80 per cent of rural counties do not have a public bus service and 40 per cent of the rural population have no form of public transport (Woods 2005). Mattson (2014) reported that only 13 per cent of US individuals living in rural neighbourhoods have available public transport and public transport's mode share in rural areas was 0.9 per cent. Poorer residents in the UK and Europe are more likely to be public transport dependent; only one-third of French rural communities and one-half of English rural communities have a daily bus service (Woods 2005). According to the UK National Travel Survey, in 2012, some 11 per cent of UK rural households were without access to a motorcar (compared to 28 per cent in urban areas) and 49 per cent in the most rural areas had a nearby regular bus service (compared to 96 per cent of urban households) (DOT 2014).

Many of the services most needed by disadvantaged rural communities and those who are less likely to have private transport, such as hospitals, welfare and benefits centres, are in distant regional towns or cities. Rural households that do not own motorcars are often those who are elderly and disabled in some form and have very limited opportunities for independent mobility. Younger people not yet of driving age also face mobility problems in rural areas and school leavers encounter mobility barriers in efforts to access opportunities for work, training and higher education. Social isolation in rural settings often has a gender bias, such as in young families with one car that is used by the male breadwinner for commuting leaving the remainder of the household without access to transport during the working day.

Despite the general barriers facing rural public transport, there are considerable differences in the types and frequency of services. As Gray *et al.* (2008, p. 113) describe: 'On the whole, "rural to urban" journey makers travelling into major cities and conurbations tend to be well served', with rural areas that are crossed by inter-urban public transport routes benefiting from this access, as do places where Park-and-Ride exchanges give access to bus and rail services. For the remainder of rural areas, providing 'rural to rural' and 'remote rural to urban' trips has found conventional public transport wanting.

Meeting the challenges of rural transport has given rise to a very great variety of national responses amongst the developed nations, and importantly, very wide differences in meeting the mobility needs of rural households. Gray *et al.* (2008, p. 111) offer the following explanation:

> A major factor in determining whether the transport network can meet the needs of journey makers in rural communities is the attitude of planners, politicians and the state, and the willingness of these actors and agencies to intervene in the rural transport market to provide a higher transport service level than the market alone would support.

In developed nations that have pursued neoliberal transport policies, rural communities experience minimal levels of public transport services as a consequence of efforts to minimize public subsidy. In the US and Australia, for instance, rural transport services have been subject to deregulation, privatisation, reduced or eliminated public subsidies, service rationalisations and performance requirements. Rural communities have experienced some of the greatest losses of public and subsidised transport following neoliberal reforms; services have tended to become restricted to profitable routes and networks become fragmented. Following neoliberal changes to the UK bus services in the 1980s and 1990s, many unprofitable services were closed or rationed resulting in the rise of rural on-demand services and community-based minibus and taxi services, supplemented by 'supermarket buses' to regional supermarkets (often to the detriment of local businesses) (Gray *et al.* 2008). In the US, scheduled intercity services that offer rural residents access to cities declined between 2005–10 so that the percentage of rural residents without access to these services increased from 7–11 (some 8.9 million people in 2010) (BTS 2013). After several years of worsening rural mobility and rising mobility inequalities, the UK government reintroduced several subsidies and incentives (to local authorities) with resulting improvements in rural mobility through the 2000s (DOT 2014).

Western European nations with exemplary rural transport systems are those that largely eschewed the neoliberal principles in transport policy. Oftentimes, much of the differences across Europe and within European nations are due to the differing willingness of local government authorities to support rural transport. Switzerland is notable in this regard, providing as it does, comprehensive rural public transport. Its success is based on more than public subsidy (indeed, its level of subsidy appears to be at the less expensive end of the continuum) and is founded on national public policy commitments to providing equitable access to mobility, in which rural services are integrated into the national transport network.

Demand-Responsive Transport

Achieving effective rural public transport in developed nations has been generally achieved through demand-responsive transport and other flexible approaches. Demand-responsive transport covers an array of services, but generally is characterised by:

- mobility services are available to the public and using smaller-capacity vehicles;
- flexible routes and schedules (as opposed to the fixed routes of conventional public transport);
- flexible destinations and origins are available for some services;
- shared rides;
- responsiveness to the needs of individual passengers.

Users in most schemes can reserve services in advance to pick-up at home or nearby; operators typically seek to combine pre-ordered trips into cost-effective routes (thereby providing a cost advantage over taxi trips). Such systems became popular in the 1970s and 1980s in many nations, but with the worldwide adoption of national and other legislation to support the mobility needs of those with disabilities unable to use conventional public transport, there was rapid growth in demand-responsive transport, notably from the 1990s onwards. Greater attention to the needs of rural travellers through this period saw demand-responsive systems more extensively employed for single journeys or to access conventional public transport. Demand-responsive transport in developed countries is, in many senses, the regulated version of informal transport in many developing nations (as described in Chapter 2).

Demand-responsive services take several different forms, including those available at public transport interchanges (with integrated ticketing and timetables), services that have specific destinations (such as employment centres and airports), networked systems of demand-responsive services and those services that are independent of public transport services. Accordingly, the markets for such services are relatively wide, covering welfare and special needs travellers, those without access to private vehicles or public transport to travellers seeking a high-value service. Most services are publicly subsidised, although some are commercially viable and serve niche markets. Accordingly, demand-responsive services can be provided by a range of providers: local governments, state government authorities, commercial entities and community transport (Laws *et al.* 2009, Enoch *et al.* 2004, Davison *et al.* 2014). A great majority of community transport provides demand-responsive services.

There are a great number and variety of these services around the world, particularly because most schemes are small scale and operate locally. Estimates of the number of these schemes are not readily available, although some national and international surveys and reviews have been conducted (e.g. Laws *et al.* 2009, Davison *et al.* 2014, Enoch *et al.* 2004). For instance, the majority of the 1,500 US rural systems are demand-responsive services (Ellis and McCollom 2009). Europe has had a number of large and successful taxi-based systems since the 1990s: Publicar and CasCar (Switzerland), Regiotaxi and Treintaxi (Netherlands), Anruf Sammel taxi (Germany) and TaxiTub (France). These systems, that are often of a larger scale and operate across regions, are within integrated regional transport systems and in some cases are national schemes with national funding. Publicar, for instance, is operated by a subsidiary of the Swiss Post Office across several cantons with 112,506 annual trips (CIT 2008). Treintaxi is a national taxi-share scheme that serves around 100 rail stations, providing some 2.2 million trips per annum (CIT 2008). At the other end of the scale are systems providing for a few hundred passenger trips annually.

In the face of on-going rural transport inequities, some advocates consider demand-responsive systems as the ideal model for meeting the needs of the disadvantaged. Mulley and Nelson (2009) consider that broader programmes of flexible transport services could considerably enhance conventional bus public transport services. On-demand rural services were advocated by the UK's Commission for Integrated Transport (CIT 2008) in emulation of successful Dutch and Swiss systems. Serious barriers and challenges remain; for instance, in rural Scotland, Velaga *et al.* (2012) found that these included small scale operations, disconnection with conventional transport modes, difficulties in serving the diversity of rural travellers and a failing to use contemporary information and communication technologies. And of course, there remains the on-going debate over what constitutes an acceptable level of public subsidy. Despite the instabilities of the sector that has constant initiation and failures of small-scale schemes, a number of experts consider that further technological developments are improving the prospects for more and more successful on-demand operations (including reducing costs, increasing route flexibility and dealing with demand fluctuations) (Ellis and McCollom 2009, Velaga *et al.* 2012).

Integrated Rural Transport and Transport Networks

On-demand services do not enjoy universal support; many schemes have failed for a variety of reasons, often related to planning and design factors (e.g. Enoch *et al.* 2006), to which can be added the usual concerns over the relatively large subsidy demand-responsive services often require. Advocates of conventional public transport have also rejected widespread demand-responsive systems with concerns that these expensive systems are inferior to well-organised public transport. Petersen and Mees (2010), amongst others, consider that greater integration of public transport services and the benefits of effective transport networks as practiced in rural and semi-rural Switzerland offers a cost-effective model for servicing such areas and a viable alternative model to demand-responsive approaches. This critique touches on an important distinction that has been underplayed in the advocacy for increased demand-responsive transport and that is it is ideally complementary to a networked public transport system and does not compete with regular public transport services. As Petersen (2009) describes of semi-rural Switzerland's Weinland region, it is possible to operate a high-quality public transport service using local bus services coordinated with the rail system, with the network designed around servicing the rail stations. In this case, the villages are relatively compact and many are situated on the larger roads, facilitating an effective networked system that provides regular hourly services for a large portion of the day and evening. Certainly, the successes of the large-scale demand responsive transport systems and the best-performing demand responsive systems share a common feature in being well integrated into wider public transport networks and being highly planned operations.

Combined Goods and Passenger Transport

Sometimes called 'postbuses', the practice of carrying goods and passengers has been long established in many countries, including the Austria, Germany, Switzerland and the UK. In the UK, mail postbuses carry over 50,000 passengers annually. These are essentially routine and daily (often with only two daily trips) mail delivery journeys on which passengers can hail the postbus and be collected (albeit usually subject to some conditions, such as traffic and safety considerations). These services offer flexibility in terms of stops, but typically have fixed routes and timetables and inward and outward journeys may take different routes. As such, while these services provide a valuable form of mobility, there are obvious limitations in the types of journeys that are practicable.

Differentiating Welfare Transport from Community-Owned Transport

In some countries, the term 'community transport' refers to a form of welfare transport or paratransit. As might be inferred from the preceding materials, community ownership of transport is not confined to providing a welfare service. Under the objectives of sustainable transport and the goals of environmental justice, there are social objectives for community-owned transport and it is easy for confusion to arise over the place of welfare mobility in the case for community ownership of transport.

At the outset, it may be best to state clearly that the function of community-owned transport is not to provide welfare mobility, although there are no reasons why it might not include such functions within its activities. Welfare mobility goes by many different labels around the world and there is quite a range of eligibility for such services. Governments and charitable organisations provide most of these services for disadvantaged groups, including the elderly, handicapped, indigenous peoples, ethnic minorities and youth. As is well known, there are many causes for people's restricted ability to access transport services, including mental and physical handicaps that restrict mobility or cognition and inadequate income. Transport systems impose barriers to use, including areas and times of poor or absent services and the illegibility of the transport system. Facing the barriers of those needing public transport, the range of trips sought by those welfare groups, and the range of welfare transport services, it follows that there are many different types of welfare transport, ranging from basic regular services, location-specific and provider-specific services, services that integrate with existing public transport, regular services and irregular services. In most developed nations, public authorities and public resources directly or indirectly support welfare transport, with considerable contributions from charities and from community volunteers, resulting in a wide variety of provider organisations. Support often comes from a variety of government agencies and typically several will be providing some support for welfare transport services, such as by departments involved in social welfare, health, ageing, education and training, rural services and community safety. Business models of these service providers very greatly, as do their cost structures.

Transport agencies are sometimes critical of welfare transport, although their concerns typically reflect the outlooks of transport providers that do not generally cater for passengers with special needs, as is required of welfare transport. Such transport-provider failings of welfare transport include the small number of patrons, limited hours of operation (particularly during evenings and weekends) and small service territories. Many services are on-demand and users may require additional assistance, so that the labour inputs to these services can be high and the number of daily trips per vehicle can be very low. Oftentimes, welfare services give the highest priority to utilitarian trips, such as to medical services, with social and leisure trips given a low priority, so that welfare transport is not a full substitute for the mobility enjoyed by the general public. Vehicle fleets providing welfare transport are often comparatively under-utilised, with vehicles having spare passenger-carrying capacity, with little institutional capacity to optimise services, and there is usually an absence of coordination of services with regular public transport. Critically, welfare transport can be the only mobility option for severely disadvantaged citizens, with all that this condition implies for the necessity of such services.

Reconciling Public and Private Models of Transport Ownership

By identifying the limits of public transport we are not seeking some neo-conservative agenda to attack all things associated with government and the collective good – far from it. Indeed, even those who reject public transport and any associated expenditure, will acknowledge that no large modern city could function effectively without some measure of public transport, as systems devoted to the private motor vehicle with minimal mass transport facilities suffer the worst effects of road congestion. Adopting the models of private ownership of public transport, as discussed above, will do little to overcome the pre-existing limits of public transport.

Community-owned transport might offer a way to fill the gap between public transport and private transport, thereby providing an alternative to these established ways of providing urban passenger mobility. Community-owned transport could combine some of the desirable features of both private transport and public transport and simultaneously offer a way to avoid the costs and limitations of both. In short, what is sought from community-owned transport is the autonomy and satisfaction of individual mobility preferences, user engagement in decision-making, greater innovation and improved environmental performance.

Much neoliberal reform of public transport has been promoted by a political agenda rather than by economic or social rationales, so that the outcomes of neoliberalism often appear to be better suited to corporate stakeholders in this process than consumers of transport services or taxpayers in general. Community-owned transport is at least as capable of delivering many of the benefits sought from neoliberal reforms, including having service providers being responsive to consumer needs and preferences, offering improved mobility services, operating with lower costs and being innovative with technology and organisational arrangements. On the flip side, community ownership offers for governments a way of providing mobility without having to deal with several of the problems of privatisation. Such avoided problems might include the need for on-going subsidy to private operators, the need for corporate oversight and monitoring of contractual agreements, the risks of fragmentation of the public transport system and the loss of economies of scale through having multiple service providers (such as for information systems and marketing).

Models for Community-Owned Transport

Community ownership can take many forms, as can communities themselves, and here we review a range of possible and actual models for community ownership. As is known from studying the social institutions developed and used to manage a vast range of commons, there are many different institutional models available. Most of these institutions were developed organically from specific cultural and commons resource circumstances, but some have been designed to suit the needs of commons management in industrial societies. A few prominent models under which there has been considerable variation in actual practices are examined here.

Publicly, there has been a vigorous debate in many nations and cities over the relative realms of public and private transport, such as arise in public policy contests over privatisation and infrastructure financing, as outlined in earlier chapters. Because community participants themselves make the decisions over community ownership, the details of these decision-making processes and discussions are largely outside the realm of public knowledge. Accordingly, this limitation circumscribes understanding what values and processes come into play in such decision-making. Similarly, full public disclosure of many institutional decisions made by corporations and governments involving transport is often not forthcoming; if anything, the increasing corporatisation and privatisation in public transport has shifted much information about these decisions behind the curtain of commercial confidentiality.

How can there be Alternatives to Governments and Corporations in Transport Ownership?

Given the relative rarity of alternative ways of owning and managing transport services outside the state and corporate models that have prevailed throughout the era of modern transport, it could be imagined that these alternatives are simply anomalies. A counter argument is that the rarity of alternative ownership models does not reflect economic and cultural factors as it does the influence of political interests, major stakeholders and established institutions (that are subject to path dependency) in the management of a particular set of commons resources and services. One of the features of contemporary debates over the roles of states and markets is the use of property rights, with proponents of neoliberalism and conservative politics in particular framing social and economic policy debates in terms of property rights. Looking at the question of property rights and the rationales for government intervention in transport can help in understanding the case for alternatives to government and corporate ownership.

In the world's wealthier nations and increasingly amongst the wealthier in the developing world, vehicle ownership by individuals is commonplace; indeed, following housing, car ownership is typically the most expensive household asset in households. Yet, the transport system itself depends almost exclusively on public infrastructure and public institutions. Transport is a perfect exemplar of the problems of collective goods; without centralised ownership, modern transport becomes impossible. Optimal performance of these systems requires a high degree of order, so there needs to be uniform control over infrastructure, systems of use and control, safety, standard practices and other such matters. This can be understood in different ways. Setting up the first rail lines for horse-drawn trams in cities required the agreement of city governments, for example, because the use of such public space required the permission of a legitimate authority that represented the interests of the populace. In a sense, all urban transport is based on this model; urban transport requires use of commons property to a high degree because it needs to traverse urban space, needs to coerce the cooperation of other stakeholders for a group of exclusive rights to a range of public resources and needs a complex set of institutions to manage the transport system that impinges on other institutions. We might then turn around the popular advocacy for corporations under neoliberalism and suggest that it is in fact quite difficult and complicated to find a place for private firms and private ownership in urban transport systems; many special allowances are necessary to have private ownership in urban transport because conceptually and figuratively, transport cuts across so many boundaries of common property.

Private ownership of vehicles and rolling stock is straightforward and exclusive property rights over such assets offers few problems for owners or the state, but vehicles are merely decorative artefacts without the transport infrastructure and the systems that make the vehicles useful. Under neoliberalism, there has been an increase in private ownership in transport systems, both in transport infrastructure and in public transport. These changes suggest some fluidity between public and private ownership, so that it is possible to have private ownership of infrastructure, public transport services, and transport administration, management and planning. At first glance, this appears to validate the claims that private property can be readily applied to transport systems in their entirety. By implication, privatisation of urban transport could continue to completion.

There are, however, two very important provisos that apply to the private ownership in urban transport and in sum, suggest that the property rights of individuals and firms are quite conditional. First, in the design of privatisation of public assets and services, the state maintains a degree of control in the public interest and sets and monitors the performance of the privatised entities to ensure this interest is protected. States maintain this interest because the types of market failure that required state intervention in the first place usually persist and the design of privatisation schemes centre on accommodations to the persistent issues of market failure. For example, where a privatised service is a natural monopoly, the state will act to ensure control over price-setting and access rules by the private owners to prevent the abuse of market power through price gouging. Secondly, because of the complexity of state involvement in urban transport, there are a great many niche activities that can be readily undertaken by private firms without running into major market failure problems. Oftentimes large public transport authorities, notably public railways, operated under charters to generate certain rates of return or had earnings targets set by governments and they would use various enterprises to generate profits, such as retailing in railway stations and renting of advertising space in stations and rolling stock.

A simple guide to this phenomenon is that privatisation occurs where there are potential profits for firms; state enterprise activities that do not offer private gains are not privatised. Such essentially commercial activities have been privatised without incurring any great problems of market failure that the state would need to address. Significantly, although privatisation suggests fluidity between private and public ownership, this is illusionary. Behind the newly privatised enterprise sits the powers of the state because privatisation does not alter the character of market failure that requires governments to act in the public interest. Privatisation and corporatisation can occur in activities where the state's role in responding to market failure is not undermined.

Social Enterprises

There has been considerable interest in the social welfare and philanthropic sectors in recent years in a business model known as a social enterprise. Pinning down exactly what constitutes a social enterprise is proving difficult. Some definitions are particularly broad, such as offered by Ridley-Duff and Bull (2011, p. 1): 'an umbrella term for any form of organisation that innovates or trades for a social purpose'. Under this umbrella is any community-based enterprise (owned by employees or service users) undertaking activities in the marketplace and investing its profits into social and environmental causes, including cooperatives, not-for-profit businesses, credit unions, community-owned businesses, employee-owned businesses, charities and micro-finance institutions.

However, for the purposes of this volume it may be helpful to stick with some of the narrower and more specific applications that can serve as particular models for community ownership. A social enterprise is a type of business, usually under community ownership of some sort, operating for profit that is directed towards a social or environmental purpose. Unlike conventional firms, social enterprises do not seek profit as an end in itself, but as a means to an end. As such, their goals are set autonomously and not imposed by states or corporations. Essentially, these enterprises are not public sector, nor are they conventional forms of private enterprise. Traditional for-profit firms, firms with active social responsibility practices and socially responsible firms are not, therefore, social enterprises.

Social enterprises operate in every economic sector and cover manufacturing, retailing and wholesaling, and trade in local, regional and wider markets. They vary widely in scale. Resources are derived from fee-for-services, income, in-kind contributions and from paid and volunteer labour. These enterprises can have either an external orientation for beneficiaries, aiming to benefit the general public or an external group; alternatively, their orientation can be internal to benefit enterprise members, or be a mix of both (Ridley-Duff and Bull 2011). Some social enterprises operate in one field and direct their profits to an entirely different field, such as a thrift shop whose owners support a health-related cause in another country. In this volume the interest is in those models where the enterprise is directly engaged in providing some aspect of mobility.

International growth in social enterprises is due to several factors. A number of sources suggest that this growth is a response to the failings of markets and the welfare state, but others argue that this social economy corrects the market forces promoting economic efficiency and state interventions for social justice, so that it constitutes a 'third sector' (OECD 1999). Some find that market failures (covering externalities, information asymmetries and failures to take equity and redistribution into account) and the state's problems, such as budget limits, lack of responsiveness and bureaucracy, are causes (OECD 1999). A 'social economy' was described by the OECD (1999) in the following way:

> By aiming to revitalise economic and social life by effectively implementing new mechanisms based on a spirit of solidarity and sharing, the social economy offers alternative forms of management, power sharing, transmission of knowledge and even evaluation of results of projects that it runs or to which it contributes.
>
> (OECD 1999, p. 12)

Sector growth, therefore, is due to changes in the welfare state, market failures, increasing demand for social and community services, and new incentives for employment and cooperation between economic and social actors (OECD 1999).

Benefits and Roles

Usually founded in particular locations in response to particular issues, concerns or welfare needs, the membership of social enterprises comprises local networks or groups with both local knowledge and motivations to address particular social needs. Closeness between the participants and local issues is a notable strength of social enterprises. Several other benefits have been identified, including the productive outputs of social enterprises, as opposed to philanthropic entities that basically re-distribute resources without adding to overall economic output. Further, the economic risks taken by social enterprises are not borne by governments or corporations. As such, social enterprises enjoy considerable autonomy and this is partly the appeal of addressing social and environmental issues through these market-based institutions.

Applicability to Community Ownership

Social enterprise models have several characteristics applicable to community-owned transport. Even if understood narrowly, social enterprises offer many choices and models for communities to organise themselves so that almost regardless of local circumstances, there likely will be at least one suitable enterprise model worth considering. As circumstances change, communities may need to alter aspects of their organisation and social enterprises can offer this flexibility if designed appropriately, as opposed to forms of government and corporate ownership where organisation change can be difficult due to factors as legal restrictions or opposition from corporate owners or shareholders. Although the administrative burdens may be excessive for very small enterprises, business models can be used at a wide range of scales and can often be scaled up or down.

Business models, including those not-for-profit, can provide a strong framework for organising community activity, such as accounting principles, record keeping and reporting, including offering a range of legal protections and conferring a certain legitimacy from the perspective of the state and general community. Other organisational responsibilities, such as health and safety, insurance and liability, workplace and other employment conditions, may also be clearer under a business identity than for more casual and ad hoc organisations. If a community-owned transport enterprise own assets of some value, then a sound legal footing offers a measure of assurance and confidence for all participants and provides some legal protection for any personal or corporate investors. Agreements or arrangements with governments and corporations are also likely to be easier with social enterprises with a legal identity than with more informal groups, and if community-owned transport is formally linked to existing public transport then such agreements may be an important aspect of the social enterprises' functions.

Profit, or more specifically, for whom such profit accrues, is the feature that distinguishes social enterprises from corporate ownership. Freed of the need to make profits according to corporate objectives, such as rate of return requirements, social enterprises have a wider array of choices in organising and operating their enterprises. Taxation may be mandatory for all enterprises in some jurisdictions and there is no *a priori* reason why social enterprises should be exempt from taxation since the enterprises seek to enjoy the benefits of being within the corporate realm, although in some places, because of the charitable character of community-owned transport, social enterprises may be tax exempt. Certainly in many jurisdictions, not-for-profit enterprises have tax exempt status.

Competition is another feature that distinguishes corporate ownership from community ownership organised as a social enterprise. For private enterprises in open markets, there is competition for business and the logic of capitalism is founded on the activities of firms that compete for limited resources and compete for sales of their goods and services. Social enterprises may or may not compete with other firms, either with private firms or other social enterprises, because they may be operating in markets that have no or limited commercial potential.

Consumers in a Changing Economy

In the dot.com boom of the 1990s, there was much excitement over the prospect of the Internet and associated technologies changing the face of capitalism. Certainly, a new generation of businesses and entrepreneurs emerged and old businesses scrambled to keep ahead of the newly emergent competitors. Given the subsequent dot.com crash, at least a portion of this excitement and investment was overly optimistic – although now, some two decades or so later, the impact of the new information and communication technologies has been transformative in many aspects of social and economic life. There was a new jargon to describe this activity, including 'b2b' and 'b2c' models, referring to 'business-to-business' transactions and 'business-to-customer' activities. These phrases seem less used now, but they are a guide to an interesting observation: there was little commercial interest or awareness of the commercial potential of transactions within the group being classified as 'consumers'; no one bothered much with the idea of 'consumer-to-consumer' business.

Without doubt, this was a significant oversight, as the growth of the peer-to-peer business sector (as it has come to be called) has been very rapid and in many ways is one of the Internet's greatest economic achievements. This effect is something well beyond the falling cost of information sharing, as the growth of the sector results from people identifying products and services that they can sell that previously were not particularly practicable or appealing. Some of the world's largest and fastest-growing enterprises in recent times have not been established businesses selling to other businesses or to customers, but have been those businesses that facilitate peer-to-peer activities, such as eBay, Airbnb and Uber (as described in Chapter 4).

There is nothing very special about most of these activities in the peer-to-peer business sector; the services and products are generally of small scale and the income generated by the participants is similarly modest. It is the successful companies that provide the services that are noteworthy because of the sheer scale of the small transactions by the participants that they facilitate. Optimists in the dot.com boom might have lost much of their investments, but their failing was perhaps one of unfortunate timing rather than underestimating the extent to which the Internet and information technologies would transform social and commercial life. Very few saw the potential for how the most mundane of commercial activities could provide the basis for massive commercial growth – and those that did have surely reaped the rewards. A large-scale peer-to-peer business sector now appears to be a fact of life. While the global Internet open to the public is not that old (circa the mid-1990s, for the sake of argument), sufficient time has passed, so that a new generation of adults in the western world have not known a world without it.

So might the idea of community-owned transport benefit from the new economy that utilises the Internet and has found a place for peer-to-peer businesses? Information exchange allows individuals and communities to learn from each other, to find out what works and to learn from what failed. As such, the experiences of one community enable another to develop their own transport services more readily. More specifically, technical knowledge can be shared and those with specialist skills are able to disseminate their knowledge and insights more readily. This is potentially particularly important for the pioneers in any area, for by definition, they otherwise have no other source of experience from which to draw. Organisational management amongst a scattered community is readily facilitated through the Internet; in short, there are means and opportunities for group organisation and information sharing without the necessity of physical co-location. Indeed, some consider that the Internet allows uncoordinated groups to exceed the achievements of formal institutions (because of lower transaction costs) as the Internet promotes sharing, cooperation and other forms of collective action. There are also choices as to how open or closed such organisations wish to be. Although it may be contrary to some interpretations of the spirit of community organisation, the Internet can offer a low-cost path to marketing the ethos, ideas and services of community groups to a wider pool of potential members and supporters.

While there are many opportunities for individuals and communities for cultural enrichment and for income generation in this digital era, there are clearly costs to be borne and many contingent risks. Accordingly, some caution is needed in viewing the future in the digital era and already many troubling issues have arisen, such as the loss of privacy in the digital realm, theft through the digital realm, communication between criminals and terrorists and the 'digital divide' that further widens income and social disparity. Author Lee Siegal (2008, p. 3) observed that: 'criticism of the Internet's shortcomings, risks, and perils has been silenced, or ignored, or stigmatized'. It would be an error, therefore, to assume that the Internet's continued growth in scale and influence will be entirely beneficial. Indeed, there is no need to endorse the vision of the Internet's promoters at all when observing the extent of the influence that the Internet currently enjoys and will no doubt continue to enjoy for some time yet. Despite concerns, the growth of peer-to-peer business activity is significant, has been essential for the new economy businesses in carsharing, ridesharing and bikesharing and could well be directed towards the cause of community ownership of transport.

Invention and Innovation and Mobility Services

Such is the divergence of views about these concepts that it can be hard to initiate any discourse; for sheer convenience, invention is taken as the creation of something new and innovation, the implementation of inventions into practice. Dealing with community-owned transport, where the realms of change might range from technology (such as new applications for mobile devices) to institutional initiatives (such as new forms of enterprises or institutions to provide mobility servicers), there is an overlap between invention and innovation. Because this work is interested in both the concept and potential and actual applications of community ownership of transport, the focus in this volume is largely towards innovation.

Writings about the new economy sectors – invariably featuring information technologies – positively promote the concept of 'innovation'. Older sectors of the economy, such as transport, are more likely to be upbraided in such writings for an absence of innovation. There has been, of course, much innovation in urban transport throughout its past and contemporary history and some particular places, such as Singapore, are renowned for their innovations in public transport. Goldman and Gorman (2006), for example, wrote of four innovative directions offering to ground the quest for sustainable transport within transport systems that they label as new mobility, city logistics, intelligent systems and liveability. A considerable number of volumes have highlighted transport innovations from around the world, such as Cervero (1998), Hoogma *et al.* (2005) and Schiller *et al.* (2010). Indeed, those places with exemplary practices could hardly have achieved such outcomes without someone having led the way the way with a strong commitment to risk-taking and experimentation.

We need to acknowledge here that 'innovation' is a rather fraught term, lacking any real consensus as to its meaning, let alone how it might be achieved. Economics dominated the early consideration of the idea, with it being associated with new or existing knowledge being applied to create new associations or developments that applied to knowledge, resources, processes and products usually with the goal of net economic gain or commercial advantage. In this field, innovation also applied to organisations, structures and institutions. Traditionally, the sources of innovation were the corporate sector, public sector, institutes of research and higher education. More recently, innovation is seen as having a much wider field of participants and might readily involve product consumers, not-for-profit groups, the volunteer sector, independent experts and hobbyists. Innovation involves, therefore, changes of sufficient scale and significance to be noteworthy and brings forth substantial benefits over and above those achieved through routine refinement of existing practices and arrangements.

Reasons for the general lack of innovation in transport by the major state and private stakeholders are often the unintended consequence of regulation and control over the transport market. Monopoly providers are highly regulated by government, but not all problems of monopolies are resolved through regulation, and large transport corporations tend to follow established approaches to their operations, investments and organisational arrangements. Regulatory controls tend to evolve more slowly than operating conditions, so that regulatory controls stifle change and responsiveness in transport service providers. Bureaucracies regulating monopolies are prone to 'regulatory capture' by the firms they regulate, a condition that promotes regulatory passivity and compliance rather than promoting higher performance. Investment in rolling stock and infrastructure by monopolies tends towards uniformity, rather than seeking optimisation through flexibility and differentiation. Partly such caution is a function of scale, but the absence of competition also fosters settings antithetical to innovation. Barriers to new entrants to existing urban transport markets mean that even rising operating costs and demands for new services are often insufficient to encourage innovation. However, de-regulation and other neoliberal reforms (such as privatisation of public sector services and public subsidy reduction), do not axiomatically lead to increased innovation.

Reasons for innovating are as varied as the motivations for any human activity, but from the perspective of social sciences, of particular interest are innovations producing substantial social change (whether intentionally or accidentally). In the modern era, the idea of widespread and major social change resulting from technological innovation is commonplace, given the range of technological advancements since the Industrial Revolution. Some of these changes add to existing milieus, but for the most part it is now expected that major innovations are both disruptive and destructive of existing practices (i.e. Joseph Schumpeter's excessively-quoted concept of 'creative destruction'). Oftentimes, the wider social effects of innovation are largely unpredictable and although much writing on innovation deals with the positive benefits, the counter-balancing reckonings of any losses are often underplayed or missing.

One older model of innovation is based on inventions taken to markets by entrepreneurs with a business model for applying and marketing the innovation, with eventual success or failure having as much or more to do with the entrepreneurs' decisions and capacities as with the quality of the invention. Innovation by community-owned transport enterprises is less likely to follow this model and is more likely to result from new solutions to existing problems, so that success is less about market acceptance of change and more about the efficacy of the proposed solution. Indeed, there are new ideas about innovation that seek to exploit the capacities of consumers, so that new ideas do not only emanate outwards from firms but have origins in the marketplace that are harvested and sorted by firms. Views such as these also take into account a wider set of contributors with different perspectives, interests, training and expertise that can all contribute to innovation, whereas older models of innovation emphasised specialised expertise within a closed group. Another newer perspective on innovation emphasises the dynamic character of successful innovation; conventional views of innovation concerned perfecting the elements of the innovation process before being taken to market, but increasingly, success is thought to be more likely by responding dynamically to external conditions with innovations that are flexible and malleable in accordance with market signals (sometimes called 'open innovation').

For private enterprise, there were good reasons to shift away from traditional models of innovation. A pressing limitation was the phenomenon that firms that got to market first with an innovation often fared worse than those who waited until the initial problems of applying the innovation were resolved before following suit. This 'wait-and-see' strategy was often rewarded in practice. Countering this effect was the rise of small innovative companies (often linked to the rise of venture capital firms in the late twentieth century) that came to rival and sometimes overtake large established firms with large research and development capacities; this effect has been far less evident in the transport sector, however.

Community ownership of transport is, of course, an innovation in owning transport, managing local transport services, financing local mobility and in other ways. Community-owned transport is likely to spur innovation in the mobility services; community-owned transport also seems well positioned to benefit from these new views on innovation. There are several reasons for this receptiveness to innovation and change in community-owned transport: these organisations are comprised of those using the services provided, they are not burdened by having a commercial profit incentive and are able to experiment, can adopt organisational arrangements to suit innovation strategies, can learn and adopt from other groups and can innovate across a range of policy, managerial and operational areas. Examples of the types of innovation could include service types, service planning, business models, management, communicating with users, associations with public and private transport and links with other kinds of services. There will also be some types of innovation in transport, however, that will only be suitable for large organisations in public and private transport.

Not-for-Profit Carsharing

There is a considerable range of enterprises and experiences with not-for-profit carsharing, and while most attention has been on the growth of the larger corporate carsharing operations, there has been a considerable growth in the number of small, local carsharing enterprises. However, not only have there been considerable success by the not-for-profit enterprises but the most successful have rivalled their commercial counterparts. One of the better-known of these success stories is City Carshare in San Francisco. Started in 2001, City Carshare was founded by a small group of transport activists and began with a fleet of one dozen VW Beetles. This non-profit organisation now has 400 vehicles; 45 per cent are either all-electric, plug-in hybrid or hybrid vehicles. Overall, the vehicle fleet covers a range of vehicle types, including cars, wagons and vans (City carshare n.d.). Operating from 240 locations in the San Francisco Bay area, the service has 15,000-odd members (City Carshare, n.d.). Savings to members compared to the costs of car ownership and use and estimated by City Carshare at US$8,400 annually. Several of its innovations include a programme for wheelchair-accessible vehicle services, an agreement with a business park with a special fleet of electric vehicles for short distance trips, providing low-income residents with subsidized rates and working with land developers to put carsharing into new residential developments.

In the UK, community-owned carsharing schemes are known as community car clubs, the largest of which is the Moray Carshare in Findhorn and Forres in Scotland with around 70 members and 11 cars (Carplus 2015). There has been considerable growth in these carsharing enterprises in recent years, but there is no consolidated database available to provide an accurate account; most of these carsharing schemes are small scale (with two to ten vehicles) and there is a considerable variation in their operations, funding, membership arrangements and vehicle types. This range includes carsharing schemes with one vehicle in a rural village, all-electric vehicles fleets, all volunteer clubs, schemes with employees and services that include journey sharing. There are several firms providing support services to these carshare schemes, typically offering attractive insurance cover, online booking services, helplines, billing and vehicle maintenance. Carplus is the accreditation body for UK car clubs and it serves to promote carsharing and provide assistance and advice to community groups in establishing a car club.

Whether a cooperative run along corporate lines qualifies as community ownership might be questioned by some, but one of the world's largest carshare schemes operates on this model and it is a noteworthy example of its type. Switzerland's Mobility Carsharing is a member-owned cooperative and is the largest and longest-established carshare operator in Europe; some 4 per cent of Switzerland's drivers are members. There are 55,000 cooperative members, but non-members can use the service and in 2014, Mobility had 120,300 customers (46 per cent were Mobility members) (Mobility Cooperative 2015). Their fleet comprises 2,700 vehicles at 1,400 sites in 500 locations across the nation; there are some 186 employees (Mobility Cooperative 2015). Efforts have been made to have a vehicle fleet with lower emissions; in 2014, the average emissions were

95 grams of CO_2 per kilometre (Mobility Cooperative 2015). It is also expanding its operations to include bicycle sharing and other mobility options.

Community Ownership and the Neoliberal State

Despite the proposition that the era of open-ended neoliberalism has generally concluded (Docherty *et al.* 2004), the fact remains that Western nations and a good many developing nations have public policy settings and institutions that are results of the era of neoliberalism. Welfare states have not so much as disappeared, rather the essential goals and instruments of such states have very much been incorporated and subsumed into a broader neoliberal enterprise of the state. Although it may appear that neoliberalism has become less popular as it attracts less commentary and debate, the explanation could well be the opposite; many of the ideas and practices of neoliberalism have become the 'new normal' in many nations and that the limits of neoliberal reforms have been set by voting populations seeking to retain the remaining vestiges of welfare state institutions and services.

Market opportunities arising from the withdrawal of state services are not necessarily attractive to corporations, and nowhere is this clearer than in urban transport. Corporations engaged in service provision to governments or taking up franchise or full privatisation opportunities in public transport do not undertake these because of the potential profits on offer from assuming full responsibility for the services. Nearly all of the world's public transport services operate at a loss (even when past infrastructure and rolling stock investments are excluded). Private firms typically have no interest in relieving governments of such burdens and seek to engage only in selected activities where operating profits are possible. Governments, therefore, have to partition the potentially profitable components of the public transport systems in order to be able to attract private suppliers, while continuing to subsidise the remainder of the system with public funds. Indeed, it is not unusual for the contracted service itself to continue to receive some public subsidy, an arrangement that creates risks of rent seeking by corporate interests (Mees 2010).

From the perspective of promoting community ownership, neoliberalism may be a neutral influence. Neither governments pursuing more market-oriented practices in public transport nor corporations wish to service locations with low commercial prospects. In other words, those places and times with low and/or intermittent demand, focused on short trips in local areas, are of low commercial potential. Governments seeking to reduce expenditures will seek ways to reduce services to such areas. Consequently, the case for community ownership becomes stronger the greater the adoption of neoliberal approaches to mobility provision by the state. Governments and corporations can hardly raise objections to community initiatives in mobility services in markets that are deemed sub-commercial. Financial success by community-owned transport would invariably excite the interests of private firms that might seek to acquire such businesses, whilst states might become interested in opportunities to capture part of the value of community enterprises in monetary form to contribute to the public purse. In short, while the supporters of neoliberalism have no ideological grounds for supporting community ownership of transport, conversely, there are no compelling reasons why community-owned enterprises could not find ways to flourish in their own ways within jurisdictions following neoliberal approaches to transport.

What Community-Ownership Transport Offers: A Summary

In summary, the case for community-owned transport is based on the roles it can play in urban mobility:

- *Increasing equity in mobility access and use*: Community-owned transport can provide transport services in locations and times where existing public and private transport does not meet user needs.
- *Increasing the use and viability of public transport*: By connecting more travellers with public transport, community-owned transport extends the reach of public transport services (including resolution of the 'first mile/last mile' problem), and thereby increases overall mobility for the community through public transport networks.
- *Viable alternative to private motor vehicle use*: Where community-owned mobility services can be substituted for private car use and motorised transport there will be the potential for reducing the environmental costs of transport.
- *Capacity for change and growth that is not dependent on public finance*: Community-owned transport does not depend on public sources for capital or operating costs, so that its development is not subject to political decision-making by city and/or state governments concerning public transport and private transport infrastructure.
- *Local mobility needs and preferences can be satisfied*: Community-owned transport is designed to meet local community mobility needs in ways decided by those communities.
- *Mobility services are directly accountable to community owners*: Community ownership can engage users in decisions over management of the service, policy, and mobility services.
- *Local services can be more innovative and creative than large, centralised operations*: Small scale and local groups can find new ways to develop new technologies, practices and organisations suited to local and specific needs.

8 How Community-Owned Transport Addresses Unsustainable Transport

Sustainable transport has been the subject of a great many books, articles and research papers and there are many public policies and corporate initiatives to reduce the environmental and social impacts of urban transport. Comparatively little attention has been directed towards activities at the local scale, with most given to individuals, cities, companies and governments; community ownership is not considered in the conventional discourse. This chapter makes a case for the role of community ownership in efforts to move towards sustainable transport and begins by suggesting that community ownership can make a unique contribution.

A Unique Contribution

Community-owned transport can contribute to environmentally sustainable transport in a number of ways. Reducing the environmental, social and economic costs of conventional urban transport systems entails actions across a broad spectrum. Approaches for achieving this transformation extend across many aspects of contemporary industrial society, covering technological change (such as involving vehicles and fuels), information and communication technologies, social and behavioural change, public policy, corporate activity, operating and managing transport systems, land use and land planning, urban land management, economic and market-based measures, advertising and marketing campaigns. Within these categories of activity are myriad choices, options and approaches. As discussed in previous chapters, the challenges of sustainable transport extend across the different scales of governance, involve the public, private and civic sectors and entail a range of public policy realms simultaneously. Nearly all policy documents and scholarly works on sustainable transport adopt approaches that either specialise in some aspects, such as technology change and public policy reform, or adopt a broad-brush and multi-faced array of initiatives. Taking a comprehensive approach to the problem of sustainable transport within a single jurisdiction is not possible. Oftentimes, these works take as a starting point the concerns of a particular set of stakeholders and identifying the strategies to meet the needs of that group. As a result, there can be widely differing recommendations on how to achieve sustainable transport simply because stakeholders, such as government agencies, international development NGOs, communities and corporations, for example, all inhabit essentially different realms.

What nearly all of the existing approaches to sustainable transport have in common is that they are largely technocratic in their outlook, an understandable feature given the history of the transport sector, and that they are managerial in approach. In other words, the strategies for sustainable transport are typically classic 'top-down' recommendations. How transport needs and demands are understood in this model is through information and data collected and analysed on transport activities, often used to forecast future demand. This 'predict and provide' model has been subject to considerable criticism for, amongst other things, its supply side orientation that serves to promote, rather than manage, demand (Owens 1995). Even the proffered solutions and alternatives to 'predict and provide', such as backcasting, share the same reliance on technocratic and managerial responses to transport policy. Where community-owned transport differs radically from these established approaches to sustainable transport policy is that it seeks solutions based in employing available resources and technologies, using local knowledge and represents a 'bottom up' style of problem solving based on meeting local and specific needs.

Despite such differences between community-owned transport and traditional transport approaches to sustainable transport, there are also substantial areas of common ground. Several of the current approaches to sustainable transport can be promoted by community-owned transport, including increasing access to mobility services, increasing the levels of mobility services, reducing the levels of private car use and increasing public and active transport, reducing car-dependency, changing urban land use planning and use and promoting access to destinations that reduce total demand for travel. Many of these changes are underway around the world and community-owned transport can contribute to these efforts, as well as offer some unique contributions; community-owned transport can also be a disruptive influence on established urban transport.

Reducing Mobility Disadvantage

Much popular and policy discussion over social welfare is concerned with the 'access and equity' disadvantages faced by vulnerable sectors of the community in obtaining and using necessary goods and services, including employment, welfare services, education, health, housing, communications, food, water and energy. Often overlooked in these considerations is mobility. Yet urban transport and personal mobility can determine access to these goods and services; inadequate access to mobility services can be a major contributor in inequity in accessing services to meet human needs. In developed nations, the age of automobility has proved especially problematic in providing universal access to urban mobility. With the decline of public transport that occurred as a result of automobility in many OECD nations, caused in large part by those who could afford private motorcars switching to private transport en masse, those of lower economic status were left to depend on public transport and the rationale for mass public transport began to feature two arguments linked to automobility. First, it was argued that public transport now performed a welfare function, picking up those who did not drive. Second, was the lesser theme that public transport use was alleviating potential road congestion by lowering private driving levels.

Such arguments did not help the cause of mobility access. As cities grew through suburban development in the post-war era, private motorcar travel became the default planning assumption for mobility in the developed world. Gaps developed between total urban mobility needs and the levels of service that were universally available; urban society began to divide between private car users and public transport users, with surprisingly little overlap. This meant that those relying on public transport were now depending on services of declining quality, for as patronage fell, levels of service were also dragged down. Public subsidy of public transport invariably increased and came under scrutiny, especially so after the rise of neoliberalism.

There are exceptions, of course, there always are. In this case, the cities that had relatively little suburban development tended not to follow mass automobility. Prominent in this group are the successful Asian nations that have reached the levels of prosperity so as to enter the category of developed nations, such as Japan, Singapore and South Korea. Through the twentieth century there were also a number of European cities (such as in Switzerland) that retained tramways and urban development was shaped by such transport choices. Cities of these nations tended to build large public transport networks for mass mobility and automobility became the mobility choice of elites. Subsequently, public transport users in such places do not have the different class and socioeconomic profiles as the non-Asian OECD group.

Those without access to motorcars in prosperous developed-world cities were minority groups: the young, the elderly, the recent immigrants, the impoverished and the socially marginalised. For example, the UK's Social Exclusion Unit noted that in a 'car-centred society' those without a car and on low incomes had a diminished ability to travel, as walking became less suitable as destinations became more distant and dispersed: rising bus fares reduced their access to public transport; changing services undermined public confidence in service reliability, road congestion reduced public transport punctuality; and public transport services received less financial support from governments, reducing services (Social Exclusion Unit 2003). Collectively, although the number suffering transport disadvantage was numerous, there were few instances where such minorities could be seen collectively or be able to summon collective political power. Prosperous and growing middle-class suburbs demanded new roads and freeways from their elected representatives and were successful at the macro-scale. There was, however, no universal access under automobility and the era of automobility was marked by a social demarcation. Car users enjoyed independent mobility underpinned by public subsidy that rarely drew comment and public transport users and those with few mobility opportunities were usually only conspicuous in public and political debate as welfare recipients whose public funding support was deserving of close scrutiny.

It may be too sweeping a generalisation, but the governments of the developed world that followed automobility have generally not been particularly concerned with access and equity to mobility until recent decades. One set of political and public policy responses was to treat the problem of inadequate access to mobility through established social welfare mechanisms and provide welfare mobility services to those of special need. Philanthropic organisations and established charities have also played a major role in welfare mobility. Of course, there are those with particular mental and physical handicaps who are unable to use public transport and require special services, but these are not the bulk of those without access to mobility.

Rarely was the problem considered as a consequence of the assumptions and practices of public policies that supported automobility; elected representatives and public sector officials typically failed to recognise the systemic failings in urban mobility as outcomes of transport policy decisions. Somehow, the lessons that might have been drawn from cities with strong public transport systems that were providing relatively equitable mobility (compared to automobile cities) were not learned. Welfare recipients and their inability to afford private transport by motorcar was deemed to be the problem, rather than the inherent iniquity created by automobility.

More recently in the current century's revival of public transport in the developed world, there has been some attention given to what the advantages of increased mass mobility might bring to widening social accessibility to mobility. Again, in the developed world the target users for new public transport services are inevitably the middle- and upper-classes. In cities with marked differences in income (i.e. most places in the developed world) and marked differences in the levels of public transport service (again, most places), it can be difficult to ensure that public transport services are distributed equitably because such services tend to attract higher land valuations. Even in cities with high levels of car use and which are extreme exemplars of automobility, public transport stops and stations boost local property values, so that over time, the poor are priced out of locations with better public transport access as those who are wealthier move in.

Community-owned transport addresses mobility disadvantage by improving access and availability to transport by providing local transport, giving access to established public transport, and improving the capacity of walking and cycling to connect to existing transport networks. Typically, we would expect community-owned transport to offer considerably lower mobility costs for users than private vehicle use. These services create larger transport networks offering greater connectivity, flexibility in travel plans, more travel plan options and a wider range of mode options. Local transport services can encourage greater concentration of local services, especially if these are promoted by sympathetic land use planning, so that the trips' lengths to reach services are reduced. Local deliveries and outreach services also have greater viability within smaller service areas, enabling government and charitable providers to serve welfare recipients more effectively and economically. Greater localisation of retail, commercial, employment, education and recreational services can bring proportionally greater benefits to those of lower socioeconomic status as the costs, both financial and of time, are lower for local journeys compared to longer trips. Such strategies can supplement the usual welfare approaches to subsidising transport, such as concession prices or vouchers for public transport.

Increasing equity in access to mobility implies those with increased mobility opportunities through community-owned transport will be travelling more and those with abundant mobility choices will not be travelling less. Consequently, the environmental costs of transport will increase with more overall mobility if everything else stays the same, putting the equity and environmental goals in opposition. It follows that community-owned transport must seek the least harmful options for providing mobility from the existing technology choices if it is to minimise its environmental impacts. More widely, however, community-owned transport needs to become a more routine and substantial component of the transport system so that it can play its part in providing alternatives to private car use, which is the primary source of environmental harm within the transport system. Although this problem might now be novel to transport (i.e. the trade-off between equity and environmental protection), it has been a part of sustainable development discourse for many years.

Local transport can also become a part of the fabric of local institutions and community-based activities for social and recreational purposes. These relationships can serve an important function in providing support for those in need of welfare services without having to engage formal welfare services. Social relationships in defined communities enable ready exchange about services and opportunities, spreading information and knowledge through informal channels. Government authorities and public transport providers spend considerable amounts on advertising, marketing campaigns and information systems in efforts to raise awareness of available transport services. Reaching disadvantaged groups through such means is often problematic; local conversations may be just as effective in spreading awareness of transport systems and choices. Sharing of goods and services between community members, including transport, is fostered within local communities so that the quality of life for those in need of welfare support is improved without great cost to the people involved and in ways that do not involve governments or corporations.

Pollution Reduction

We can depict all the urban transport options as being ranked from the least polluting to the most polluting, using an index such as the greenhouse gas emissions per kilometre per person. Most polluting will be single person car journeys, next will be cars with passengers, then there will be various public transport trips, and below that walking and cycling, and lastly, avoided trips. Within this hierarchy, many further divisions are possible, such as by the type of car and the loading levels of public transport. Essentially, the goal of pollution reduction from transport is to move as many trips down the 'ladder' of pollution as possible, the most significant of which is to shift the majority of trips from private cars to public transport and active transport.

Although a widespread policy approach to motor vehicle pollution is the advocacy of less polluting engines and energy sources, the prospects for rapidly reducing pollution through these means is poor, with most of the technology gains from fuel efficiency improvements having being achieved. As Moriarty and Honnery (2008) have shown, the prospects for success in the alternative fuels and engines approach has not and will not achieve widespread diffusion and uptake in the immediate future and the greenhouse savings of many measures are suspect (such as from biofuels, hydrogen and electric vehicles).

As such, having reduced private car use to an absolute minimum, the major policy options for reducing transport pollutions are public transport, active transport (i.e. walking and cycling), and travel demand management (i.e. travelling shorter distances and travelling less often), such as achieved by land use policy and planning. Public transport and mass transport have an essential role to play in the quest for environmentally sustainable transport. Simply put, the only viable options for achieving widespread passenger mobility with very low social and environmental costs in the immediate future are public transport and active transport. Private car use will no doubt continue for a long time to come, but this pathway will not lead to sustainable mobility because there are no current, viable, and mature technology choices that can deliver this goal in the immediate future. There may be options for the longer term, such as electric cars when the national electricity grids are based on renewable energy, but from the perspective of global environmental decay, such solutions will be too late.

One condition must be stated clearly: community-owned transport denotes an alternative way of owning and managing passenger transport and does not necessarily deliver sustainable transport. Community-owned transport can, however, contribute significantly to sustainable transport and on these grounds can be pursued as a way to improve the environmental (and social) performance of passenger transport.

We can measure environmental impacts of transport in a myriad of ways depending on what information we have and what we want to know. Such measurements are complicated and involve tricky questions of where to draw the boundary, for instance, whether to adopt lifecycle accounting, well-to-wheels assessments, direct and indirect emission accounting and vehicles and infrastructure assessments. Because of the importance of global climate change, assessments of the greenhouse gas emissions of different transport choices has become the most popular yardstick of recent times of how the different transport modes perform. For the sake of relative simplicity, a simple hierarchy of pollution by the different modes of transport can be established on the basis of the direct greenhouse gas emissions produced by vehicles on a per capita per distance measure.

Community-owned transport services conform to several key strategies for sustainable urban transport. Shifting mobility down the ladder of emissions per passenger kilometre is a primary objective of sustainable transport, the most important aspect of which is to change the individual journey in motor vehicles to a less polluting choice, such as taking a journey by public transport or active transport. Community-owned transport can cater for such a change as it can cater for such a substitution of the private motor vehicle.

Community Engagement

Community engagement embraces a range of activities involving the community in mobility planning, decision-making, organisation and delivery. Relating these activities to sustainable transport may seem at odds with much of the accounts of sustainable transport that rarely has a place for the mechanisms of transport planning, community engagement or the ethics of public policy concerned with procedural justice (as opposed to the ethics of the consequences of policy). Elaborate reasoning need not be evoked here, for a simple justification should suffice. At the most fundamental level sustainable transport is more than a system of mobility that satisfies a series of performance metrics and it is more than an assembly of technologies and practices to produce the desired environmental performance (noting that this performance is invariably measured in terms of doing less damage to environmental and social values and adding value to economic variables, rather than restoring lost values and minimising economic harms). Sustainable transport is, in its ideal form, the manifestation and expression of the tenets of environmental justice. As such, the goals of social justice are as central to sustainable transport as are the goals of reducing the ecological losses that it causes to occur.

Economic Benefits

Change and innovation often faces the challenge that the economic costs of change will be excessive, but in the case of community-owned transport there are a number of economic benefits from the aforementioned reductions in the environmental costs of urban transport, both from reduced or avoided costs and from new sources of income and pecuniary benefit. Several economic benefits arise from reducing the economic costs of transport pollution that could result from reducing private car use, increasing mass transport use, and greater walking and cycling associated with community-owned transport. Broadly, transport's contribution to urban greenhouse gas emissions could be reduced and some of the costs of future climate change impacts and adaptations lowered.

More specifically and locally, lower transport pollution will result in lower health costs associated with transport. By reducing private car use and promoting greater public transport, community-owned transport shifts mobility to lower risk modes and reduce the not inconsiderable economic costs of road trauma. Community-owned transport can promote more local journeys, thereby reducing total distance travelled and lowering risk exposure. Promoting walking and cycling will confer net health benefits. Many of the health costs incurred by governments, corporations and households from preventable illness and excessive morbidity and mortality arising from the sedentary lifestyles of car-dependent societies and from the effects of road trauma are preventable. Cardiovascular disease, chronic obstructive pulmonary disease, cancers and diabetes – the four major chronic health categories in developed nations – are all associated with sedentary lifestyles, making this a major cause of death and illness. Community-owned transport promotes greater active and public transport use, and the lower exposure to road trauma thereby reduces the total economic costs of mobility.

A major source of avoided costs to the public sector is that community-owned transport offers a way of increasing mobility without the need for major road infrastructure expenditure by public authorities. Community-owned transport relies on the sunk investments in transport systems, but offers a way to provide mobility without adding to these costs to the same extent as private motor car use. It is worth remembering that even though governments have found alternative financing strategies for major transport investments, such as public–private partnerships, Build–Own–Operate–Transfer schemes and franchising, invariably, the financial benefits to the state are traded off against the disadvantages of shifting cost recovery mechanisms into the corporate sector, such as public accountability, loss of flexibility and the need for additional government oversight. Community-owned transport relies on the sunk investments in transport systems but offers a way to provide mobility without adding to these costs to the same extent as private motor car use.

Complementary Benefits and the 'Virtuous Circle'

Reforming transportation faces many barriers, including that of believing in the intractability of certain barriers. Reformers in transport must face the entrenched views of experienced experts, typically with backgrounds in engineering and economics. By necessity, both education and working practices in these professions are cautious, conservative and tend to regard innovation from other fields with suspicion, a characteristic of path-dependent systems. These remarks are not intended as an attack on these esteemed professions that dominate transport planning and management, but we also need to be plainspoken about the barriers to reform. Yet even when the professions come on board, the influence of vested interests, biases in political processes and institutions and other barriers are evident. For example, transport economists have long advocated the use of congestion pricing as a means of reducing or managing road congestion, and although it has been taken up in a number of cities, it remains controversial and difficult to get through political processes. Indeed, it appears that such measures are only achieved with an exceptional 'alignment of the planets'.

Urban transport systems are complex systems of many different elements, so that fundamentally changing the character of these systems requires actions in a wide array of realms: individual and communal behaviour, vehicle and energy source, technologies, private–public–active transport modes, transport system management, public and private investment decisions, economic measures and economic market signals, urban planning and design, use of urban spaces, education and training, and research and development. Community-owned transport can contribute to the goals of environmentally sustainable transport across a number of strategies; it is not offered as a panacea.

Indeed, the areas where community-owned transport can be effective are not those that have received the greatest public policy attention and private sector investment, for these have been in strategies for cleaner vehicle and energy sources (fuels, alternative fuels and energy sources [e.g. electric vehicles, bio-fuels and hydrogen], new vehicles engines and drive trains and measures to improve vehicle energy efficiency); pricing mechanisms to promote behaviour change (such as toll roads, road congestion charging and road time-of-use charging); information systems to improve road traffic efficiency and 'greener' vehicle operation (such as real time information on road conditions, responsive road system management and automated vehicle operating systems) and various economic measures, notably road use pricing.

Community-owned transport offers a 'bottom-up' response to the challenges of sustainable transport. As such, it is in contrast to the more common aspects of public policy and private sector investment in sustainable transport initiatives. Both the existing public policy-led approaches and bottom-up approaches have a role to play in sustainable transport. Sustainable transport's goals stand a better chance of being achieved and achieved more quickly if the special role that community-owned transport can play is more widely recognised by public authorities and the broader community.

9 Likely Objections to Community Ownership

Community ownership is far from the usual ways of providing services in modern societies, especially in the richer and more economically developed nations and cities and as such, there will be many objections to its application to providing urban transport systems. However, community-owned transport is not enforced communal ownership or a regime that prevents private ownership of the means of mobility. Rather, it represents a choice and offers an alternative way of thinking about transport systems that seeks to supplement existing urban transport systems. Two types of potential objections are the pragmatic and the ideological/political. In practice, as we know, many public policy debates use pragmatic arguments as proxies for political values and preferences, protecting vested interests and unexamined social prejudices. Here, we concentrate on pragmatic objections to community ownership of transport and in the next chapter, deal with the political economy of such ownership, where aspects of the political debate over community ownership are examined.

It is possible to speculate on the likely pragmatic objections based on the tenets of orthodox transport policy and planning. As stated earlier, transport policy is strongly based in the disciplines of engineering and economics, a foundation that has equipped the sector to progress rapidly in many aspects, such as mechanical efficiency, organising complex public transport systems and identifying low-cost options. Countering such strengths are many weaknesses, such as those described in the creation of mobility systems with a poor environmental record and failings in addressing access and equity in mobility provision. These weaknesses include the tendency to scientism, strengthening of path dependency and institutional inertia and preferences for conservatism and predictability in social institutions. In reviewing these objections, some counter arguments are advanced and ways in which such objections might be reduced or nullified are considered.

Increasing Fragmentation of Transport Systems

Urban growth tends to complicate the transport task and as transport systems evolve in scale, scope and mix of components, the task of control and management increases in difficulty. Technological and social change creates new opportunities and demands for mobility. Institutions managing and operating urban transport also grow and multiply, so that over time there is a trend towards greater institutional complexity. As a result, there are more agencies and separate jurisdictions, with increasing specialisations and functions. Policy-makers and transport providers' tasks grow in scope, detail and complexity over time and a greater number of stakeholders are involved from government, business, non-government organisations, research groups and lobbying and advocacy groups in decision-making processes. Consequently, there arises one of the conundrums of governance, namely the inexorable rise of fragmentation of the overall transport system, of its control systems, of agencies and institutions, of accountability mechanisms and other features. Not surprisingly, growing urban transport systems inevitably face a growing scale of challenges and problems, something plainly obvious from the history of urban transport.

A prominent response in transport research and advocacy in recent years is that of greater integration in transport systems, a broad concept that has been applied at various scales: operations, institutions and land use and transport planning. Complicating this matter is that the integration means quite different things at these different scales. At the operational level, the goal of integration is to allow for seamless trips across complex transport networks, so that differences in modes, systems, owners, operators and locations are bridged in ways that ensure that journeys are not disrupted regardless of circumstances. Integrating operations between different modes (such as between buses and urban rail), different operators and owners and separate systems invariably requires changes in governance in some form, such as new operating agreements, new institutions and new public policies. Such integration is sought and achieved through top-down and centralised authority models of reform.

Such networked systems greatly enhance the attractiveness of public transport, given that transfers are a major source of consumer dissatisfaction. In the absence of effective public transport networks, many experts think that it will be very difficult to lure car drivers across to public transport. Working out ways to make integration appealing to state and corporate stakeholders can be difficult and regulatory intervention has often been necessary, together with invested public funds, for while there are efficiency gains on offer, there are also potentially significant expenditures required.

Producing such seamless journeys requires bringing together many aspects of these systems, including such elements as coordinating timetables and interconnecting services, single ticketing systems, unified information, advisory and trip planning services and public transport interchanges. Achieving such integration involves a number of technological and technical activities, institutional arrangements, legal and regulatory instruments, technical specialists and expert knowledge. For systems that developed in fragmented ways, and that is typically the case regardless of city size, integration is a challenging and often confounding task.

Adding a new layer of disparate, local and independent transport providers in the form of community-owned transport could result in greater fragmentation of urban transport and further inhibit network creation. As opposed to centralised and authoritarian models, such community-based models provide opportunities for bottom-up approaches to decision-making and service delivery. Community-owned transport could add to the complexity and potential fragmentation of urban transport through:

- New local community services that will shift mobility patterns and add a greater burden to existing public transport services.
- Demand for coordination between community-owned transport and public transport.
- Transport system planning by central government agencies and transport service providers that seek to take community-owned transport into account.
- Infrastructure requirements to facilitate the needs created by community-owned transport.
- Additional or revised regulatory systems for the special needs of community-owned transport, such as if government authorities wish to register such services.
- Community-owned transport stakeholders that seek engagement in wider transport policy and planning processes.

For transport systems where community-owned transport becomes popular, none of these changes to urban transport seem particularly unlikely given what we know of the existing transport systems. Many integration advocates would be likely to reject this development, such is the particular importance attached to the cause of public transport integration.

One response to such objections is that community-owned transport services could be integrated into the existing public transport systems, as have other changes in the history of urban transport. Put simply, local community-based services could be coordinated with existing public transport systems. How difficult this task becomes depends on how the community-owned transport system operates, its primary objectives and the interests and organisation of existing public transport services. It is likely that many users of community-owned transport will have public transport stops and stations as a popular destination and as such, there is an obvious incentive for some types of integration between public transport and community-owned transport. Because community-owned transport responds to the users' needs, connectivity with other transport services is likely to be a high priority for those wishing to access the wider urban transport networks. Rather than these services fragmenting the public transport network, it is more likely that the role of community-owned transport will be to increase integration (and in ways that do not involve an extra burden of coordination on existing operators). Arguably, community-owned transport has a far greater incentive for integration than existing operators in public transport and faces less barriers in doing so, as their operations are so much simpler and have only to consider local community interests, presumably in a relatively small geographic area.

Another response is slightly more provocative, and potentially controversial, namely that integrating community-owned transport is not as important as it is for entire public transport systems, so that a weakly connected or unconnected community-owned transport may be acceptable in light of the other benefits of community ownership. Many ad hoc, informal, intermittent and local transport services in cities around the world that are operating effectively without coordination with the larger transport system and with few apparent costs. This phenomenon illustrates a principle of organisational theory in which there are diminishing benefits for additional management efforts past a certain level, so that attempts at controlling an entire system can be counter-productive and may turn out to lower overall performance (i.e. the law of diminishing returns). A more intellectually ambitious response to the problems that integration seeks to address is to suggest that allowing for a degree of self-organisation can be more effective than a comprehensive management system that cannot ever have sufficient information and means of control to achieve optimal control of a complex system. Therefore, a good case can be made for sensibly limiting the scope of management ambitions in efforts to control complex and diverse systems.

Seen from the transport users' perspective, as opposed to those of transport advocates or transport system operators, integration can only be valued when an individual traveller is able to access public transport and then, only as much as the system itself can provide the mobility sought. Where public transport systems are poorly or highly unevenly developed, effective integration does little for the prospects of areas and times without decent services. Providing mobility in the existing gaps is where community-owned transport services will be most valued, so that criticising community ownership on the grounds that it is not ideally integrated with other services is akin to claiming that no transport service is to be preferred over poorly connected services. Again, from the perspective of some stakeholders in public transport, such reasoning is effectively (if never explicitly) what underpins the design and operation of the system, otherwise how could there be acceptance of unmet mobility demands by governments and corporations?

Community-owned transport will not rival public transport in passenger volumes any time soon and never could in large cities under foreseeable circumstances. Accordingly, the task of integration will remain squarely within the domain of the relevant authorities and corporations and how quickly and effectively integration is achieved in any city will have little to do with community-owned transport. This is not to argue that integration should not be a goal for all urban transport providers, but to consider community-owned transport as a barrier to integration is essentially a straw man argument.

Inefficiency of Small and Local Services

For the owners and operators of public transport systems and urban transport systems in general, efficiency is the Holy Grail. This is at least the case because of the esteem this concept holds in the two disciples that shape the thinking and decisions of the urban transport sector, that of engineering and economics. In turn, it is the professionals trained in these disciplines that have shaped transport bureaucracies and who, in turn, influence governments' perspectives and priorities. Oftentimes, the goal of efficiency is so essential and central that it governs decision-making without even being identified as such and when it is advanced as a rationale for decisions, it is rarely defined, explained or justified. Efficiency is more than a trope or figure of speech in this discourse and it assumes the role of a professional article of faith. As such, when something is deemed to be efficient in a transport system, it is being endowed with the highest qualities, while that is deemed inefficient is problematic, undesirable and a failure that needs to be rectified.

If we examine the annual reports of transport organisations, the language of efficiency appears regularly. It also appears frequently in public discourse about transport systems. Use of the concept, however, does not always refer to any concrete measures, but because the concept carries the identity of science and rationality, its users often seek to imbue arguments and statements with the cache of this thinking. In looking at the charge that community-owned transport might be inefficient, we need to consider with care exactly what such a charge might mean so as not to be swept up in an empty semantic debate.

Defending the goal of efficiency, proponents adopting the engineering approach may claim a consensus over the concept, proposing the familiar story that efficiency refers to maximising the gains (as outputs) from a specific set of expenditures (as inputs). Any outcome that fails to achieve the nominal ratio of output from an input is inefficient, because there has been wasted input and potential forgone output. Economics, similarly with its interest in the allocation of resources within markets seeks out those transactions that maximise the gains from any exchange that has the lowest transaction costs. What should be immediately apparent here, even from within the perspective of these two disciplines, is that efficiency is a highly abstracted and conditional concept. In both these approaches, careful quantification is usually a necessity of efficiency assessments, as is the delineation of the elements to be measured. To underscore this point, a public transport operator aiming to operate a beautiful service would likely be met by public derision or confusion, but one aiming to have an efficient operation would pass unnoticed, yet both goals are arguably similarly abstract.

Efficiency is abstracted in three obvious ways: there is something produced from a process that is identified as an input, something that is identified as an output, and there is a process that links these phenomena and that this process can be assessed by an account of these inputs and outputs. In the social sciences, these abstractions are the starting point of a number of concerns that give rise to a questioning of the validity of efficiency assessments, of the acceptability of efficiency assessments and of the broader politics of efficiency assessment.

How then might efficiency assessments be invalid as an exercise in performance assessment? To begin, there are potential problems arising from measuring inputs and outputs. In an array of outputs, say from a public transport system, there will be some that are amenable to measurement and on which data has been collected, just as there will be benefits that are difficult to measure and for which we do not have a good understanding, let alone much in the way of data. As a result, what is assessed about transport system performance reflects the things that can be measured; it is a short step to claim that these measures are the most important for managing the system. But in a mix of the known and unknown, how can it be claimed that what is known is necessarily more important than what is unknown? Such a critique is similar to the insights from the concept of bounded rationality, which, generally speaking, suggests that rationality only makes sense within a set of accepted pre-conceptions related to the boundary of the system involved. Limits are, by necessity, an artificial de-limiting of the complexities of the world, but by implication if we include other interests in an assessment of rationality, then our existing propositions about efficiency can no longer be assumed to hold.

On the second type of concern with efficiency, we can draw some strong insights from the American writer on civilisation, Lewis Mumford (1967), who argued that efficiency makes for a dangerous social goal in itself, as it mistakes means for ends. It matters little, he argued, if undesirable outcomes are being produced efficiently. In transport systems, claims of high efficiency are of little merit or desirability if the outcomes that are being produced are unimportant.

Third, there is a broader set of concerns about rationality as a social phenomenon, of which efficiency is a hallmark. This is quite different from rejecting efficiency because it can be abused as a tool, but takes as a starting point the proposition that efficiency is axiomatically loaded with its own politics and is far from being a neutral assessment whose impacts depend on specific circumstances. One stream of concerns deals with the promotion of technological determinism that holds that society inexorably benefits from technological developments and, further, that science and technology offer a blueprint for the organisation of society. Such arguments have a long history in utopian writings, with notable contributions from Saint-Simon. Aspirations for a rational society have drawn much criticism and certainly, many of those promoting rational societies come with a number of very unpleasant political values. These criticisms include elitism, authoritarianism, centralisation of state power, the privileges of the scientific and technical classes, corporatism, bureaucratisation of decision-making, narrowness of values and technological determinism. What does this mean for the goal of efficiency in social systems? A ready example is the claimed greater efficiency of dictatorial decision-making over participatory and democratic decision-making. A short list would include a questioning of the basic distribution of power amongst decision-makers, a questioning of what is subject to efficiency assessments, a questioning of whom and what is subject to such efficiency assessments and a question of how efficiency is to be assessed.

Opponents of community-owned transport might point to several obvious practical ways in which it would be inefficient in comparison to existing public transport systems. Such an objection can be expected because the history of public transport is one that has so strongly and consistently presented evidence as to what is called ‘increasing returns to scale’. In other words, there are benefits enjoyed by bigger systems that are out of reach for smaller systems, so operators of urban public transport can expect better results from one large system than several smaller systems. ‘Economies of scale’ is a general term to describe the benefits of increasing the scale of an economic activity; specifically, it refers to the effect that as the level of production increases, the unit cost per unit produced is reduced (as identified by Adam Smith and other founders of economics). Such an effect can be applied to individual enterprises, industrial plants or entire economic sectors. (Its opposite is the diseconomy of scale, whereby increasing scale increases the costs of each unit produced.)

Although there are myriad reasons for economies of scale, there are a number of often-cited factors. Technical and technological factors include the opportunities that larger outfits have to invest in newer and higher-performing equipment, thereby increasing productivity at reduced costs. Experience and skills in larger enterprises can be more specialised, with the resulting expertise leading to better management and decision-making. Larger firms also enjoy the benefits of making larger purchases of inputs, such as bulk buying and also using market power to purchase inputs at advantageous prices. Some functions can be scaled up with little additional cost and some costs are fixed regardless of output levels, such as marketing and some administrative costs – factors that give larger firms a competitive advantage over smaller rivals. Certain opportunities only become available to larger operations, such as access to certain kinds of finance or the influencing of regulators and lawmakers through lobbying. Historically, larger firms often have shown greater financial resilience than smaller firms and survive periods of economic recession through strategies such as product diversification, cost savings and staff redundancies.

Given that urban transport is historically dominated by large service providing enterprises for mass mobility, then it will be argued that community-owned transport will be inefficient because of its small scale. Two ideas being co-joined in this argument can be questioned. Firstly, community ownership is assumed to be small-scale, but arguably the concept can be scale neutral – there could be small, medium or large transport services owned communally. In practice, these systems are small, but in theory, there is no reason why community ownership could not rival existing public and private systems. Certainly, if we take social enterprises as the institutional model, then it can be seen that these can be of any size, including quite large enterprises.

Secondly, the argument that small community enterprises will be inefficient is built on the proposition that the existing state and private public transport systems are efficient by virtue of their large size. Immediately, this evokes the vagueness of the term 'efficiency', but even if we take some simple economic performance yardsticks, it is clear that there is a considerable range of performance levels of these systems around the world. As is known, only very few public transport systems operate at a commercial profit; many urban public transport systems have struggled with insolvency and large operating losses over the years. On the basis of such experience, it is hard to see much credence in the case that being large of scale axiomatically confers a measure of efficiency in operations and management. If no transport experts advocate that simply increasing the scale of existing public or private public transport companies will automatically increase their economic performance, then the argument that small transport enterprises are inefficient is open to question.

There is also a counter argument to be considered: small enterprises can enjoy advantages over larger rivals, namely, the benefits arising from diseconomies of scale occurring in the so-called niche markets. There is a reasonably well-developed set of theories and case studies from the business literature on the virtues of niche markets. Mass markets can be met through mass production, but not all demands are satisfied as consumers may have demands for products and services with greater prestige and exclusivity, greater environmental or social worth or have special needs. Niche markets are differentiated from mainstream markets by product/service type and qualities and market features, such as the demographic attributes of consumers. Providers for niche markets are usually small and specialised firms servicing a market with a tight set of demographic characteristics.

One way in which public transport can be amenable to niche markets is that public transport providers are sensitive to the costs of increasing services. Although public transport is built on the principle of economies of scale, it would seem that many such providers consider that the limits of falling long-run costs have been reached. Economically, large transport providers have no financial incentive to chase additional passengers outside the existing services if this costs more than it earns. This is essentially the stage that has been reached in developed nation cities with public transport systems; they are designed to harvest the bulk of passenger needs through mass provision of services; capturing extra passengers runs afoul of the diseconomies of scale. Small, community-owned transport services do not seek mass markets and are not designed to benefit from economies of scale – they are servicing consumers that large operators do not want – making the economies of scale objection to community ownership largely irrelevant, as they are for mostly niche market operators.

Efficiency holds a special place in contemporary industrial society. A charge of being 'inefficient' is a condemnation, which in shorthand implies needless expense, but justifying this charge in urban transport services can be a difficult task given the complications of transport markets. Further, (economic) efficiency is a narrow criterion to apply to services, such as urban passenger transport, in which service providers have an array of economic, social and environmental obligations. This leads to such questions as: how important is economic efficiency in the broader scheme of life? For governments, corporations, and households, avoiding wasting money is a self-evident goal (indeed, Max Weber [2001]) proposes in *The Protestant Ethic and the Spirit of Capitalism* that this goal is pursued with religious zeal). Yet, even for corporations, and certainly for governments and households, economic goals are traded off against myriad other considerations. There can be no easy generalisations on this point, but perhaps a simple point will suffice: Ultimately, individuals and households will place their own value on mobility. Local communities are not likely to be bound by the economic calculations of public transport providers as the arbiters of economic rationality. Indeed, why should a government, corporation or think-tank decide on what makes economic sense for community groups? If the model for community-owned transport is the social enterprise, then clearly, these enterprises make their own financial decisions and the goal of economic efficiency is a matter for them to consider in their own ways. There are good reasons for suggesting that the economic efficiency of the transport system as a whole is of little importance to them.

Undermining of Public Transport and Welfare Systems

Associated with the concern that the rise of local transport services would further fragment public transport and disrupt the goals and benefits of a networked system is the potential to undermine public transport in other ways. By providing new services with community-owned transport, existing services could be deprived of current and potential customers. Furthermore, community-owned transport through such self-organising efforts, could create circumstances in which state and private interests would have a rationale to reduce their support of existing public transport. Additionally, these interests could rationalise against further extensions to the existing public transport system on the grounds that the community should take up the challenge of providing new services. At the outset, the framing of this objection to community ownership confines the community to the role of a consumer of transport and restricts the opportunities to produce mobility services to states and appointed private firms.

Such concerns raise some interesting issues about the place of the citizen in capitalist democracies in regard to public transport. Certainly, transport planning professionals do not appear antithetical to the public realm; indeed, much contemporary education of such professionals lauds (in theory) engaging the public in aspects of the planning processes. Many progressive public transport theorists and planners have promoted the ideal of community involvement in many aspects of planning and operations (as collaborative planning); many of the oft-cited exemplars of leading public transport cities have been lauded for their use of community engagement. Schiller *et al.* (2010, p. 300), for example, state: 'Many successful transformations of cities or aspects of them began as citizen efforts that ended in political and institutional changes. Often such efforts spawned activists who became accomplished experts and brought new perspectives and creative energy to their cities.'

Planners' rhetorical enthusiasm for community involvement and other aspects of deliberative democracy is often somewhat tempered in practice, with expectations for democratic engagement not being met (Granberg and Åström 2010). Having identified the value of collaborative planning with communities, Schiller *et al.* (2010, p. 301) go on to sound a warning note: 'The best-laid plans of political and community leaders could easily "go awry" without the enthusiastic support and technical skills of career government officials.' No doubt such concerns are genuine, but they also express a clear professional and public divide with the implication that public transport is too important to allow citizens to get too close to the levers and buttons of planning and operations.

For those familiar with the debate over community involvement in public policy these arguments will be recognisable. It has been a long time since Sherry Arnstein (1969) published the 'ladder of participation', but its basic construct has a timeless quality. In contemporary democracies, the low rungs of the participation ladder have achieved regular institutional responses with the type of token impacts as predicted by Arnstein. Genuine citizen participation and the idea of power sharing between governments and citizens remains, for the most part, not much more likely now in the developed nations of the world than it was when Arnstein made those observations. Arguably, in places where neoliberal policies for public transport have been given full reign, the likelihood of community participation having any significant impacts is now lower than when transport was owned and managed by government bureaucracies. Commercial operators and many public agencies trumpet their responsiveness to consumer preferences, but this is a commercial imperative that serves to underscore the political powerlessness of citizens using these transport services.

That community-owned transport might undermine specialist welfare transport appears as a serious concern, given the importance and fragility of such services for those with few other options. Welfare transport services are supplied directly by government agencies, by private operators under contract or other agreement with governments or by volunteer or not-for-profit organisations. Users of welfare transport are those unable to access private vehicle transport or regular public transport services (or, more specifically, have needs that are not satisfied by conventional transport) and are eligible for welfare transport. Providing that the service standards required of welfare transport are satisfied, then the immediate needs of its users can be met by any number of different providers, as shown by the current systems around the world where there are a multiplicity of service models. A source of concern may be rooted in circumstances where the opportunity to use community-owned transport might be taken by state authorities to remove support from welfare transport, taking advantage of these circumstances for shifting the fiscal burdens from the public sector onto welfare recipients or community groups. Such concerns would seem, however, most appropriately directed at governments and the machinery of public policy formulation rather than at community organisations meeting local needs. Public policy is formulated by governments (and their agencies) and, in democracies, are accountable to electorates, so that shifts in government support of public transport need to be determined through governmental processes; community owned transport entities cannot be held responsible for policy changes made outside their purview.

People Don't Like Buses

For a variety of reasons, it is likely that a growth in community-owned transport will have a major role for small or medium-sized buses. Promoting greater bus use will provoke the criticisms of this transport choice, many of which have a good grounding from surveys of public opinion in many nations and cities over a long period. Some of these concerns are indeed valid barriers to greater bus use, whilst others are less relevant to community-owned transport users or are the outcome of operational choices rather than inherent flaws of bus travel.

Slow trips in buses are major source of passenger frustration. Bus trip times are affected by many factors, so that reducing travel times can be difficult for operators. Sharing the roads with other vehicles involves buses in road congestion; solutions have typically been the declaration of dedicated bus lanes on major roads or other kinds of preferential treatment. While road congestion is outside the control of bus service operators, there are other factors that contribute to slow trips that lie largely within their own responsibilities. These factors include slow trips due to lengthy journeys from indirect routes that weave across cities in search of passengers and the usual close spacing of bus stops. Passenger boarding and disembarking also adds to delays, especially during busy times and where drivers also sell and/or check passengers' tickets. A range of options is available to address such issues, but for a variety of reasons (such as cost), operators may not avail themselves of these potential solutions.

Compared to private vehicle use, buses can be inconvenient. Bus services and their larger networks can be difficult to use, with little information on routes, destinations and timetables available at bus stops. Further, service frequency is often irregular and changing throughout the day and night, with further changes between weekdays and weekends. Coordination with other bus services and other public transport modes also tends to be uneven, creating delays between services. Making 'joined-up' trips can be further complicated by the physical separation of the stops and stations. Comprehensive and legible information systems on a network basis are often lacking and many cities have services supplied by competing operators, adding to prevailing confusion. As noted above, to attract passengers, buses need to service the passengers' sought-after destinations; the less direct the trip of each passenger, the greater the degree of inconvenience.

Social stigma attached to buses is a part of the problem. There is actually no record that UK Prime Minister Margaret Thatcher ever said that anyone over 30 years old (noting that this age varies by different accounts) using a bus was a failure, but this oft-repeated apocryphal phrase has not been bettered as an expression of the stigma of bus use. Switching car users to heavy rail and light rail is usually deemed to be an easier task than the mode shift from car to bus, a phenomenon tied up with the stigma of bus use. As mentioned in an earlier chapter, car manufacturers underscore the associations of car ownership and use with higher social status through explicit and implicit advertising messages. In places of high car use, buses are used by those of lower socio-economic status, students and the elderly and those without cars, whereas the rail modes tend to have a broader profile of users. Other perceptions of the relative attractiveness of the fixed rail modes over buses for car users are speed of trip, greater comfort (in vehicles and at railway stations over bus stops), convenience, reliability and predictability. Although such perceptions may be falsified by evidence and personal experiences, they remain as substantial barriers to mode shifts to buses in places with high car usage.

Bus rapid transit has been one solution to these problems by making buses' operations similar to fixed rail modes: buses run on dedicated roadways, have relatively few stops and may have express sections, offer direct routes to popular destinations and services have relatively quick travel times. Whether such services are taken up by community ownership is an open question. To date, these services are not associated with community ownership; often the rapid transit buses are a special design and of high cost. Indeed, the costs of establishing a bus rapid transit service can be very high overall (and can be similar to light rail), especially where there are special infrastructure requirements and dedicated rapid transit buses.

Many of the failings of inconvenience to travellers are not inherently symptoms of buses but of the ways bus systems are operated. Community-owned services could offer more convenient services by having effective information and advice systems, by offering services aligned with user's needs, being coordinated with other transport services, by acquiring an appreciation of the network effect in public transport operations and by servicing destination in accordance with consumers' demands. Given community ownership, aligning operations with the users' needs should be a routine and straightforward task and thereafter responding to direct feedback from the users. Offering attractive bus travel times for services sharing common roadways will be a challenge for community-owned transport as many of the influential factors will be beyond their sphere of influence. However, where a local transport service operates within a relatively small service area, travel times are likely to be short.

Problems of discomfort in bus travel can also be addressed. Bus stops and shelters are often far from comfortable and are of low amenity. Access to buses can be a major problem for those with reduced physical mobility, notably the elderly, who are often dependent on public transport. But it is the bus itself that is often the source of most complaints of discomfort; there is relatively little space for passengers, window seats have less access than aisle seats, noise levels can be high and crowding further lowers comfort. How much community-owned transport can overcome these sources of discomfort will depend, to some extent, on the efforts made by the community owners, such as building comfortable shelters or the site of stops.

Stigma against bus use, particularly in wealthy cities with high levels of private car ownership and use, is a barrier to increasing bus use, but it may be best to regard this as a challenge, rather than an intractable condition. Across the wealthy cities of Europe are places with high bus use and no apparent stigma over bus use. These places have high-quality services and the operators report high levels of customer satisfaction. On these grounds, it may well be that what may appear as a sociocultural bias against buses may be something more fluid, so that when bus services are of high quality and provide extensive services, such prejudices fall away. Indeed, it may be that these expressions of prejudice against buses are prejudices against those of certain socioeconomic groups that rely on buses (notably ethnic minorities, the poor and the elderly). In any event, it seems unlikely that individuals with an interest in participating in community-owned transport would harbour a deep prejudice against bus travel.

Impossible Idealism

Innovations in the social sphere invariably face criticism that they are idealistic and will be crushed by reality. Certainly, community-owned transport is an ideal of sorts that is set against existing practices, established beliefs and opposing vested interests. Such 'reality-based' insights can take two forms; firstly, that community-owned transport will fail because of the successful opposition of stakeholders with interests threatened by such initiatives. Secondly, that community-owned transport is an impossible ideal because of the failure of community groups who lack the capacities of self-organising involved; any successes are isolated exceptions.

There are only two groups of stakeholders with interests directly challenged by community-owned transport: governments (covering elected officials and the public service) and businesses with rival commercial interests in supplying public transport services. Of course, given that public transport is not a free market, in most instances the government and business interests are not separate, but are linked participants operating in a public transport monopoly. Business rivalry is such an obvious explanation that it seems not to require much analysis; services provided by community-owned transport represent lost opportunities for an established service provider. Is this a realistic objection, in light of the fact that community-owned transport seeks to provide services where they do not currently exist? To consider community-owned transport as robbing established operators of potential profit seems to place the basic problem at hand the wrong way around. In effect, the established operators have made their judgements over the value of providing such niche services and determined not to provide them. Further, assuming that community-owned transport services are working to the same business model as established public transport systems is patently false. It is precisely because the established service providers are not providing a service that a niche market exists. By definition, a new niche market is not a rival to existing services.

Governments are the second possible objector, exercising their role as transport market regulators. In public transport, the 'rules of the game' are the regulations that govern competition and determine, often down to a fine level of detail: who shall provide services, what services shall be provided, the performance of these services, how and when they will be provided and other factors. Political scientists have many different theories for explaining the motivation of governments and the formulating of public policy, creating a variety of ways to consider the relationship between public transport providers and governments. Some of the reasons why governments might object to community-owned transport, drawing on several different schools of thought, include:

- Regulation theorists would point to the way in which state authorities work with private companies to ensure their commercial successes. In this case, given that private companies object to community-owned transport (for whatever rationale), then governments would respond similarly, taking their lead from these companies.
- Neoliberal theorists would expect governments to object to community-owned transport because they would depict such rivalry as threatening the power and legitimacy of governments and thus prompt a reflexive response to object to such competition. This is because neoliberal perspectives depict states as always seeking opportunities to expand their powers and controls over all economic activity.
- Institutional theorists might consider the government's objections to be based on the fracturing of the domain of activities managed by government

by the community setting up rival transport institutions. They might also point to an additional administrative burden in controlling and managing community-owned transport institutions, especially if these were particularly numerous.

- Economic rationalists might point to the natural monopoly principle in public transport and offer that there is no place for rival institutions to the monopoly either owned or controlled by the state.

It is also possible to identify a set of practical objections, but most of these can probably be fitted under one the preceding groups. Such objections could include the additional costs for governments in managing community-owned transport, concerns about ensuring that health and safety standards are met according to existing laws and regulations, concerns about the qualifications and attributes of staff and volunteers, the complications of managing a transport network with a host of new service providers and the resulting loss of efficiency and orderly conduct and that existing laws and regulations do not allow for this initiative. Rather than argue out each of these objections, here we posit that there are no practical objections that cannot be overcome if governments have a will to make this occur. Significantly, in the face of change there are always objections and doubts. To consider these aforementioned items as insurmountable barriers is to exaggerate these problems and underestimate the flexibility and capabilities of the modern state. During emergency conditions governments can summon the will for great change; consider the famous 100 days of Franklin Delaware Roosevelt's New Deal in the US Congress. In the light of national and city emergencies, the sorts of routine objections raised about local reforms to transport systems seem trivial; these objections might not be trivial, but certainly all, or nearly so, can be resolved by governments exercising their authority and resources, both social and financial.

Community Ownership is Disguised Neoliberalism

Contrasting with the preceding objections that focus on aspects of community-owned transport operations and less overtly political values, such as efficiency, this chapter closes with an overtly ideological objection. In doing so, some of the themes of the next chapter dealing with the political economy of community-owned transport are presaged. Putting community-owned transport into the debate over neoliberalism yields two quite different interpretations:

- Community ownership represents an effort to counter the effects of neoliberalism.
- Community ownership is a neoliberal response.

A complication is that community ownership can span a range of different forms, so that those resembling communitarianism reject neoliberal ideals and practices, while some forms of social enterprise seem compatible with aspects of neoliberalism.

To some extent, it might be argued that within the spectrum of possible models, both of these interpretations could be readily accommodated; social enterprise models could well be a form of neoliberalism, whilst also countering big business by empowering local communities with employment, local investment of profits, being accountable to communities and by being managed and operated for community benefits.

On the one hand, such communalistic (referring broadly to direct democracy movements at the local and regional scale) endeavours are an antidote to the neoliberal dismantling of the welfare state. As states become increasingly influenced by the desire to minimise the provision of social services and apply market-based measures to determine the allocation and distribution of these services, new service gaps emerge. Communities are obliged to organise themselves and locate necessary funding and support to fill these service gaps. While governments and corporations lead the economy away from welfare functions and lessen welfare obligations, community groups follow alternative sets of values and create new institutions based on community values. In this way, communities form a bulwark to the machinations of neoliberal governments. Alternatively, bringing communalism to local transport can be interpreted as promoting and expressing neoliberal ideas as constituting an effective outsourcing of the state's responsibility for public transport to community organisations. Using social enterprises as a market-based institution is a means to manage the social costs and associated social inequality associated with the retreat of governments in providing essential services. Such local entrepreneurial efforts typify the sort of individual responsibility and self-reliance that neoliberalism celebrates. Community ownership is a sort of outsourcing and a form of privatisation that might be considered a type of libertarian municipalism.

From the broader critique of communalism as neoliberalism arises a number of theoretical concerns that can be translated into practical problems to be considered:

- Will community-owned transport favour those in society with greater material wealth and higher socioeconomic status? If so, then how far will it go towards the goal of promoting equity in urban mobility?
- What about competition between community-owned transport providers?
- Will governments act to limit the activities of community-owned transport to protect the state monopoly on public transport? Does this mean that the potential of community-owned transport is limited from the outset?
- Will state public transport take over the most successful community-owned services?
- Is community ownership elitist? Who has the education, time and skills to organise and manage such institutions?
- How might community-owned transport work in places where there is little community identity or inter-community communication?
- What does 'community' and 'local' really mean?

Answers to these questions will only emerge if community ownership becomes widespread. Ideological objections do not need to be supported by lessons from real life, but will have little credence if they are diminished by community ownership practices. Critically, there would appear to be sufficient uncertainty and countering arguments on both sides that ideological argument is insufficient to prevent greater experimentation with community ownership.

Part IV

Implications of Community-Owned Transport

10 Political Economy of Community-Owned Transport

In this chapter, the principles of political economy are used to explore the political and economic implications of community ownership of urban transport. In essence, political economy refers to the interplay of economic and political factors giving rise to social phenomena. Many of the themes of previous chapters have concerned political issues and values and here a number of these are brought together to examine several key political economy aspects of community-owned transport, including the democratisation of urban transport, the implications for state resourcing and support for public transport and the relationship between community-owned transport and public transport. This chapter opens with a framing argument that there are limitations to public and private ownership and that community ownership offers advantages for dealing with the challenges of sustainable transport.

As a sweep of history, the mechanisation of mobility represents a universalism of technology, underpinned by the applications of knowledge from economics and engineering that also assumed a universal discourse. In a metaphoric sense, the transport of modernity is represented by the laying down of tracks, something that assumed that all peoples shared a common need for mobility regardless of historical, cultural or political difference. It may be no accident that the concept of path dependency is a transport aphorism and certainly informed the public institutions for modern transport that created a web of practices that have locked in established transport systems. Tugging against the extraordinary achievements of the planners, builders and operators of modern transport systems in providing mass transport has been the persistent matter of mobility deprivations of those unable to enjoy the full benefits of mass mobility systems. Although the debate over the choices between mass private transport and mass public transport have washed back and forth, the era of automobility reached an apotheosis with the phenomenon of car dependency. Contemporary concerns over access and equity issues in transport is a particularism that draws attention to social difference and it is these differences that demark those with mobility and those left behind. As described earlier, there are distinct social contours around those with mobility disadvantage and a self-reinforcing process that makes such disadvantage a stable condition as mobility disadvantage and socio-economic disadvantage are mutually reinforcing. Distributing mobility according to economic means, gender, age, socio-economic status, cultural physical and mental capacity and other such variables hardly occurred through happenstance, but was the deliberate result of public policy and corporate choices, the behaviour of public institutions and the exercise of political values, beliefs and philosophies.

Progress, in contemporary times, has become elusive in public dialogue and has been rendered into a diminished trope; it too, has passed from the condition of universality to one of subjectivity and pluralistic interpretation. As described in Chapter 1, in the early phases of modern transport, the rise of mass transport was progress manifest. During that period of total social and economic transition, the rise of mass transport was made possible through an assortment of quite different phenomena that were simultaneously aligned – technological change, expertise, effective governance and public institutions, recognition of the need for government intervention, available funding and other factors. Underpinning these conditions were political systems that provided fairly unequivocal support for this version of progress. We are now in the struggle to achieve another transformation in passenger transport, as postulated in the Introduction and described in Chapter 3, that of sustainable transport. This is a contemporary manifestation of progress, but one of decidedly uneven development and more disturbingly, one of uncertain fulfilment. A simple explanation of the immensely complex problem of this uneven and halting transition is that the necessary conditions are not aligned. Sustainable transport is carried forward by ideologies of environmental protection and philosophical themes drawn from environmental justice. Using sustainable development approaches, governments and corporations have sought technologies, products and practices to accommodate sustainable transport within the confines of capitalist and globalised economies (see Chapter 3). As concluded in Chapter 3, these approaches have been promising but insufficiently widely adopted for success. What marks this transition as fundamentally different from those of technological transformations is that sustainable transport is without an immediate technology 'fix' – it requires a social or cultural transformation, something that marks it as a problem of political economy.

Sustainable transport presents as a problem for political economy in several ways that distinguish it from the conditions that established and promulgated automobility:

- While the goals of sustainable transport are universal, it is apparent that applying the solutions of the avoid-shift-improve kind are rooted in the particularism of local circumstances and conditions.
- As opposed to mass transport solutions, sustainable transport is founded on providing wider access to mobility and pursues equity in mobility across society as a goal.
- Bureaucratic, centralised, state-centric, professionalised and authoritarian models of transport decision-making, planning and management are at odds with the principles of wider public participation, an approach that fosters greater community engagement, provides models for change and promotes the ideal of self-determination at the community scale.

- New collaborative consumption models of mobility provision (carsharing, ridesharing and bikesharing) are innovations that have primarily occurred outside the established state and corporate stakeholders (see Chapter 4) and could be harnessed to the goals of sustainable transport.
- Corporate stakeholders and state authorities' common interests are a barrier to change, bringing tenets of neoliberal approaches into question.

Neoliberalism, as discussed below, is a fundamental element in the arrangement of influences that resists the tide of change that is sustainable transport and brings into question the political economy aspects of the respective spheres of influence and activity by states and corporations.

Limits of Government and Limits of Corporations

Limits of Government

In this neoliberal era, criticisms of government abound and political programmes for smaller government and an expanded private and corporate sector can be found in many nations, states and cities. At the outset, therefore, it needs to be stated that recognising the limits for the role of the state in mobility is not the same as opposing government. Community-owned transport sits in a two-way relationship with the state; state intervention is required to facilitate community ownership but there are aspects of community ownership that are actually or potentially in opposition to the state, in that both seek to occupy a similar role. Instituting an expanded place for community-owned transport requires, therefore, sorting out the role for government and this requires understanding the limits of government.

Politically, as might be inferred from previous chapters, support for limiting the role of government is hardly the exclusive goal of right-wing politics, as anarchism similarly seeks to reduce the intervention of the state. One of the arguments favouring the state advanced from the left is that only government has the capacity and legitimacy in limiting corporate power and controlling its activities. Because of eco-communalism's interest in independence from state and corporate intervention, from the perspective of the interests of this concept, it cannot be taken that the state is axiomatically best placed to act in the interests of the community. Several reasons have been advanced for limiting the role of the state, such as government's ties to vested interests, especially commercial interests, government's own self-interests, the problems of bureaucracies' unresponsiveness to community needs and its short-term focus that neglects long-term problems, notably environmental issues.

Greater intermingling of corporations and governments has been part of the success in maintaining economic growth within the OECD nations and the fate of elected governments has become highly dependent on maintaining a vibrant economy. This has seen the rise of new types of government support for business, new forms of state-owned corporations and state ownership of private companies and providing an array of services and support for business (such as the law, infrastructure, education and research). Governments also provide the means through which commons resources can be accessed by corporations and facilitate their privatisation. Another cost of this escalating state engagement with corporations has been a softening of the oversight and regulatory responsibilities of governments, causing an erosion of government accountability and transparency.

Limits of Corporations

Mirroring the scepticism of governments' capacities in providing services and the questioning of governments' legitimacy where there is competition with private firms is the lionisation of the corporation as performing valuable social roles in service provision. There are a great number of limitations to the utility of the role and activities of private corporations in mixed economies, however (although some of may also be applied to governments or community groups). It is not necessary to evoke anti-capitalist theories in identifying the limits of corporations.

Although corporations are bound by the legislation applicable to their operations, they are established to be accountable to their owners, who may be private individuals or public shareholders. Corporations may express their self-determined public responsibilities and these may direct their activities, but formally, corporations are not responsible to the wider community. This point is often obscured in discussions over the proper role for corporations, notably in recent times, but corporations cannot be expected to act in the interests of the community beyond their legal obligations and have a primary allegiance to their owners and their demands and expectations. Social responsibility programmes and benchmarks by corporations have become more popular in recent times, but such initiatives do not alter the essence of corporate accountability to its owners and the problem (indeed, likelihood) that the interests of the owners and the community may not coincide.

A consequence of the difference in interests between corporations and communities in its extreme form is corporate malfeasance and corporate crime. Against a backdrop of minor criminal lapses by corporations perpetuated by rouge elements are the instances of spectacular corporate scandals, where the might of a corporation is systematically organised for its own gain contrary to the law (see Salinger 2013, for example). Corporate crime is particularly important because of the scale and breadth of such activities that corporations can achieve; bringing corporations to justice challenges the law-enforcement capacities of states. In the worst instances, corporate crime is of immense scale, causing great costs to society and the environment, and creating lasting damages. Types of corporate crime include bribery and corruption, deception, misrepresentation, appropriation of common pool resources, embezzlement and other theft. Even when corporations are subject to judicial processes for law-breaking, there often remains an on-going threat of further criminal activity as legal authorities rarely have the will or means to ensure permanent changes in corporate behaviour. To this can be added the complications of corporate law enforcement within the global economy, whereby corporations can evade compliance with national laws and judicial processes.

Services and goods provided by corporations are according to the dictates of the profitability to the firm, not according the demand by consumers or the wider social utility gained through consumption. However, many goods and services might be prized by communities regardless of how essential they might be for life and welfare, their provision depends on the decisions of the corporation and the agreements they have entered into. Corporations are not expected to maximise social welfare. Wasteful resource use and environmentally destructive activities have been the hallmarks of industrial capitalism, contributing to a series of local and global environmental crises. Whether such losses are the fault of corporations given that the centrally planned economies have environmental records as lamentable as those of capitalist nations is an arguable point. Even in the wake of the environmental movement and rise of environmental awareness, corporations continue to degrade the environment in full knowledge of the impacts of their activities and for the most part, only curtailed their pollution in response to environmental law and regulation.

Limits of Public and Private Transport

Of all the debates in transport policy and planning, none is more pervasive and fundamentally divisive than the competing interests of private transport in the form of motorcar ownership and use and that of public transport. This duality reflects the familiar political left/right divide in western politics throughout the modern era, with private transport associated with neo-conservative, libertarian and market-oriented ideologies set against public transport as associated with the left wing's collectivist, state-centric and interventionist approach to markets to promote social goals. While a number of contemporary political commentators have decried the validity of such polarities (such as identifying more centralist 'third way' themes in governance), there are several ways in which private transport and public transport are diametrically opposed and irreconcilable as fundamental choices, despite whatever compromises and accommodations might be attempted.

An immediate irony marks the cause for continued *automobility*; private motorcar use depends on public infrastructure and the institutions of government. So in an absolute sense, private transport and public transport are in direct competition for government support; what one receives, the other will not. We can find other expressions of this zero sum form of competition between these tropes. Considerable urban space is dedicated to transport (and its associated activities, such as for car parking) and a direct manifestation of the competition between modes is that for road space between the needs of public and private transport for the road-based modes. At a broader scale, the competition arises in basic planning decisions over land use in the choices between car-oriented residential and commercial areas and those designed around public transport services. Both private transport and public transport have their critics. As the critique of private transport is broadly well known, we need only recount the major issues here.

Private Transport

One of the most successful marketing efforts of all time must surely be those of the motor vehicle industry to associate that most cherished ideal – individual freedom – with motoring. Yet, the liberties of private motorised mobility are highly circumscribed and the greater the number of private vehicles and car drivers there are, the more these limits become manifest. Globally, the current trajectory of the car population growth will take us to two billion vehicles by 2030 (Sperling and Gordon 2010). Such an outcome seems at odds with the implications of the peak oil thesis and it is difficult to see how mobility based on the private automobile is in any way sustainable at this scale.

Hopes that alternative fuels and energy sources will come to the rescue of the internal combustion engine and transform the existing motorcar into a less environmentally harmful commodity have carried research and development in the car industry since the mid-1970s (with considerable public funding). That we have yet to find, promulgate and bring to market a new alternative over the last 40 years is significant (given that electric vehicles are as old as the internal combustion engine). Given the immediate necessity for reducing greenhouse gas emissions and the current state of play in alternative vehicles and fuels, it is reasonable to conclude that this technology fix has not succeeded (Moriarty and Honnary 2010).

Individual car use depends on public road infrastructure funded by public expenditure (or private tollways operating under agreement with governments) and the use and maintenance costs often borne (either entirely or largely) by general revenue and taxpayers in what amounts to a major public subsidy for private motor vehicle use. As the burdens of private mobility on the public purse are shifted onto motorists, through such user pays policies as carbon taxing and congestion charging, the costs of motoring are increasing. Despite the trend towards having car users bear more of the costs of motoring, the historic and current environmental and social costs of motoring are borne more broadly (as described above). On-going car use in the general sense is not tenable under an environmentally sustainable transport system, which imposes a fundamental limit on car use around the world.

On a more prosaic level, there are other limits to car use for passenger mobility in our current circumstances. Motorcar transport systems provide socially inequitable mobility; many in lower-income groups cannot afford to own and operate a motorcar and in systems where investment in public transport has been neglected, they suffer a mobility disadvantage. Other barriers to car-based mobility include those of physical and medical handicap, the aged and children. Indeed, two key mobility issues receiving greater attention in recent years in developed nations are the reductions in children's independent mobility (such as the rise in children being driven to school) and the ageing of the baby boomer generation and the problems of driving competence, drivers licence surrender and the loss of mobility. Additionally, many of the health costs of air pollution from motorcars are disproportionally borne by those living in highly polluted areas adjacent to major roads that are typically discounted in price because of the costs of such unattractive locations. Overall urban amenity is also degraded by high concentrations of road traffic; the world's most polluted cities are characterised by such conditions.

Public Transport

A safe generalisation is that public transport is more environmentally sustainable than private motorised transport by a considerable margin, although in some instances the gap between private motor vehicles and public transport services with low passenger counts can be small (or even more polluting, such as when empty buses are compared to cars). Further, public transport typically meets a social service obligation and thereby serves to promote greater accessibility to mobility than private motorised transport. In developing countries with relatively underdeveloped public transport services, the informal (and essentially privately owned) sector provides much urban mobility. Although 'greener and fairer' than private transport, there are limits to public transport's contribution to sustainable transport and equity in access to urban mobility. These limits include that public transport does not meet all mobility needs of all potential users because of location inaccessibility, destination inaccessibility, poor connectivity between multiple services, inadequate service frequency/timing, crowding, excessive trip times, unaffordable or high cost, poor reliability, low service quality, lack of security and ease of use. Not necessarily by intention, but the fact remains that public transport provides socially and economically inequitable services and further, that the provision of services promotes inequitable economic and social benefits.

Many public transport providers are private corporations operating under contract to government entities. Such arrangements may satisfy the expectations of market advocates in prompting higher efficiency than public ownership, although this seems open to question, but such arrangements can put service obligations and corporate objectives in conflict. These conflicts might arguably be equally evident when public ownership is also managed according to corporate objectives (such as seeking nominated rates of return).

Public ownership of transport services denotes a chain of responsibility through government accountability mechanisms to the wider community, but in practice there are few jurisdictions that could make any credible claim that there is a strong system of accountability and transparency in decision-making in a practical sense, let alone many efforts to engage the wider public or service users in decision-making for the services offered. There may be community boards to represent community interests, however, for the most part it is difficult to see how these have made any significant impact and these arrangements appear to be little more than window dressing; public transport planning and operations is a distinctly autocratic and undemocratic activity. Oftentimes, consumer surveys are used to gain knowledge of user needs and views and become a part of marketing campaigns, but such consultation falls a long way short of genuine power sharing in decision-making.

Public service obligations are a primary requirement of public transport, but paradoxically, there are some downsides to this responsibility. Public transport service operators rarely have the option of withdrawing their services, which amounts to a barrier to departure. Consequently, the operators are restricted in their operating choices and will have to continue to provide services that draw down on overall potential profitability. Similarly, public transport services are usually confined to very specific territories, with similar implications for responding to market signals for profit and loss. Further, public transport companies often face relatively fixed cost structures, notably labour costs (especially when the labour force is unionised) and operating costs; these costs are relatively independent of the level of patronage.

Active Transport

Walking and cycling (i.e. active transport) have an important place in sustainable transport, especially for short-distance trips in cities. Yet, there are limits to the uses of active transport, not the least being that it does not provide for all trips. While transport planners often lump walking and cycling together on the basis that both require the exercise of human energy for mobility, in practice there is perhaps less in common between these modes than is usually taken for granted. Certainly both walking and cycling are subject to journey length limits; while individual limits will vary greatly, walking speed is often taken as a nominal average of 5 kph, cycling speed is obviously subject to a wider range, but for convenience, a nominal 15 kph can be posited. In many instances, these speeds can be greater than can be achieved in motorcars on crowded roads, but for longer trips, motorised mobility is needed. Similarly, there are physical requirements for active transport, restricting the use according to such factors as age and health status. Climatic and geographic factors also influence the speed and range of walking and cycling.

No less than any other form of modern transport, walking and cycling are dependent on suitable infrastructure, which is ideally purpose built. Of critical importance is the separation of walkers and cyclists from motorised transport, although there are those who question the need to separate cyclists from motor vehicles. Interviews with cyclists and potential cyclists around the world in many studies have consistently shown safety concerns as the highest barrier to increased cycling, logically enough given the vulnerability of cyclists to collisions with motor vehicles (Pucher and Buehler 2012, OECD/ITF 2013). Construction of new cycling paths has been shown to increase cycling rates; it seems that the benefits of on-road cycling lanes on cycling rates and collision rates is less clear-cut. Further, there is an association between increasing the numbers of cyclists and falling collision rates, known as the 'safety-in-numbers' thesis. Similarly, separating cycling from pedestrians reduces collisions. Some facilities have been created in existing urban open spaces (such as in parks and alongside rivers and canals), but creating new active transport facilities usually evokes a competition for space with other road users, such as on-road bicycle lanes or cycling paths within the roadway corridor. Such competition limits the likely expansion of active transport, especially for cycling, as politically and economically the vested interests of motor vehicle users tend to prevail over active transport advocates. Nonetheless, there has been a continual expansion in active transport facilities in the cities of the developed world, notably in western and northern Europe and in North America.

Putting the Limits of Each Mode into Perspective

As to be expected, private motorcar use, public transport and active transport are all subject to their own contrasting limitations. Each of these systems has their own advocates, however, it remains that transport systems need to satisfy many expectations, many of which are likely to be in conflict. Further, these expectations tend to belong to different jurisdictions, so that consumers may be concerned with services and costs, firms with profitability and governments with environmental performance, for example. It follows that there is no ideal that can optimise these conflicting expectations.

Public transport and private transport are mirror opposites in many respects; the strengths of one are the weakness of the other and vice versa. Much of the case made for sustainable transport is that for active transport and public transport. But even in places with good facilities for active transport and with reasonable or even excellent public transport, there can still be high levels of private car ownership and use. Even in the exemplar cities for sustainable transport in Europe and North America, achieving very low levels of greenhouse gas emissions per capita from the transport sector remains challenging. Plenty of these exemplar cities still have relatively high levels of private car use.

There are many reasons for this. Many sustainable transport advocates seem reluctant to admit that private motorised transport has some very attractive features for individual travel, especially for some categories of traveller, certain types of trips, for specific journeys and for special circumstances. Advocates of sustainable transport are understandably disinclined to acknowledge the limits of the sustainable transport modes because this risks undermining the rationale and the practice of sustainable transport. This is not to argue that a large number of people *need* their own car or motorbike, but there is a case to be made for recognising that there is a fundamental rationality about mobility decisions and in some circumstances, public and active transport are not attractive, available or even viable choices for getting around cities, regions and rural areas. Despite these limits to the use of active transport and public transport, they are the mainstays of sustainable transport and a number of the limits facing these modes can be overcome or diminished through technology choices, system design and system operations.

Convivial Society and Community-Owned Transport

A sub-theme of this volume has been to broaden the perspective of the sustainable transport debate beyond the interests, values and language of economics and engineering, while still drawing on these disciplines when needed. It has been stated earlier that one of the limitations created by the traditionally narrow disciplinary focus has been the tendency to translate urban mobility problems into forms requiring technological and economic solutions. Even when the politics of transport policy is debated, oftentimes the discourse remains framed by these dominant disciplinary outlooks. Government, business, research, academic and think-tank participants in transport debates usually seek to prove their case as being the optimal choice to satisfy either economic or technological rationales, or both. Community and advocacy groups also often argue in these same terms, fearing charges of irrelevance or ignorance should they evoke other 'irrational' values in the public debate and knowing what arguments gain the attention of public sector decision makers.

Transport policy and planning usually takes personal mobility as a strictly utilitarian function governed by rational choices that weigh and evaluate the relative merits of competing mobility options. Commuting, shopping, recreation, education, personal business and visiting are among the typical types of personal trips recognised by researchers and transport operators and managers. In this way, the matter of moving around the city can be explained as a form of consumption; surveys of travellers establish their mobility preferences and priorities. Despite the range of locations, types of travel involved and the usual variety of people covered in such surveys, it turns out that the dominant preferences are remarkably common. Convenience (accessibility to, and frequency of, services), high levels of safety, reliable services, acceptable travel times and reasonable costs are the dominant preferences in mobility. Conventionally, these preferences are a way to explain high car use in places were public transport services are comparatively poor. Certainly, this is widely taken in public transport advocacy as an article of faith with the obvious lesson being that only when public transport services can better satisfy these preferences for any specified trip will those with free choice choose public transport instead of the private car. No doubt there is some common sense in this general assertion, but given the complexity of travel behaviour and the vagaries of individual human psychology, there are grounds for caution in adopting such generalisations as these.

There are a number of assumptions underpinning this common rational model in transport planning that is essentially utilitarian. In other words, it is assumed that transport choices are the result of efforts to maximise personal utility. Missing from this view of human behaviour is recognition of social values, notably that of conviviality. Unconventional as it is, there may be merit in considering how mobility and the task of transport contribute to conviviality. What is meant by conviviality and why should it valued? In simple terms, conviviality refers to sociability, but in the hands of Ivan Illich, philosopher Wendel Berry and others, a convivial society stands for much more than mere social associations. Writing in *Harpers Magazine*, Berry canvassed the problem of social and ecological relations thus:

> As earthly creatures, we live, because we must, within natural limits, which we may describe by such names as 'earth' or 'ecosystem' or 'watershed' or

> 'place'. But as humans, we may elect to respond to this necessary placement by the self-restraints implied in neighborliness, stewardship, thrift, temperance, generosity, care, kindness, friendship, loyalty, and love.
>
> (Berry 2008)

Conviviality, then, is the antithesis of the individualistic and self-serving decision maker that is the rational ideal of conventional economics as it concerns individuals within a social group with common bonds and interests.

Under conventional transport reasoning there are other associated values and choices that reinforce the individualisation of mobility and act as barriers to thinking about collective social interests and collective solutions to mobility. In transport research, especially public transport research, a truism is that personal transport is only a derived value, meaning that people travel in order to obtain something from that trip and the trip itself is simply a means to that end; i.e. no or little value is attached to the trip itself. This is the familiar formulation of people acting as rational economic decision-makers, a founding principle of neo-classical economics and much organisational theory. Of course, car manufacturers hold to no such beliefs about mobility and have worked hard from the outset of the motorcar's invention to invest trips by car with all manner of associated values, such as luxury, prestige, sex appeal, class identity and that last refuge of a scoundrel, patriotism (Sachs 1984, Paterson 2007, Seiler 2009). Transport researchers tend to dismiss such marketing as irrelevant to the transport task and often see it as an annoying distortion of more rational signals meant to influence the logical traveller.

A contrary view is that the car manufacturers got it right in understanding that approaching mobility as a purely rational act undertaken by rationally economic beings was going to be a self-limiting view of the market for urban mobility. As an aside, it is interesting to note that motorcar advertising often features the sociability between the car passengers, so that within the vehicle friends and family are shown to be enjoying each other's company. This is clearly no accident, for in a subtle way a distinction is being drawn between the intimacy of shared car travel and the anonymity of mass transport. Mass transport is like any other service provided at the large scale. It is unavoidably designed for mass consumption, at best providing for common preferences and values but more dominantly for the convenient packing and shipping of a human cargo. It is the antithesis of individual transport and this is the conundrum of public transport in the era of private, motorised transport. Nevertheless, promoting public transport is often made out as the task of weaning travellers from the attractions of private transport essentially by promoting the compensating virtues of mass transport, something that even the sophisticated marketers of the twenty-first century have found a challenge. Car manufacturers generally have little need to promote the advantages of private and individual transport directly; but their indirect and symbolic messages attest to the pleasures of individual transport. Car travel is either an entirely private matter or undertaken with known associates, all travelling together by choice; this may be convivial, but in a highly constrained way.

This raises the question of whether public transport, as mass transport, can be considered convivial. Mass transport is largely an anonymous experience; one travels with strangers and generally the trips are kept as a functional business. Many social interactions have their origins in shared journeys to be sure, but these are a fraction of all trips and travel experiences. For the most part, at least in the English-speaking world, our trains, trams, buses and ferries are strangely devoid of social interaction; the social norm is to mind one's own business and to expect other travellers to behave in a reciprocal way. Generally, the shorter the trip and for trips in urban areas, communication between travellers is rare. At least one common recent theme in public transport marketing is the enabling of use of private communication and entertainment devices; a common sight is to enter a silent train carriage or bus with the head of every passenger wrapped in headphones or hunched over a mobile telephone. Interactions between staff and passengers seem to increasingly depersonalised in the contemporary era; public transport service staff will remain largely unknown to passengers and vice versa in larger cities, an effect no doubt increased by more flexible working hours and greater use of casual public transport staff.

Yet, it seems that either intuitively or through market research that public transport companies are conscious of the value of social conviviality. Under neoliberal influences, public transport has largely shifted from its place in welfare states to a more market-oriented perspective. Signs of this change have been a more commercial approach to service provision and public transport has invested in a stronger customer service, often led by marketing and information provision. A casual examination of the marketing literature and pictures in their annual reports usually displays their efforts in creating and promoting an impression of social conviviality. There are images of friendly bus drivers offering a smiling welcome to passengers, of elderly people sitting side-by-side talking, of tourists with backpacks being pointed the right way by a uniformed station attendant, a standing group of uniformed school students on a train and the like. Surprisingly, despite some cultural differences, it seems that public transport operators around the world all strive to create the same set of impressions and follow a marketing template. Despite these efforts, we all know this is a marketing of the ideal. Public transport has many virtues, as described in many places in this book, but to suggest that they have anything but a minor role in promoting sociability is a step into incredulity.

Under modernity, connectivity is the critical and dynamic enabler of change, so that greater mobility is a perpetual goal. It follows, therefore, that anything that adds to overall mobility is deemed desirable. Where this view falters is in assuming that all trips are of equal value and provide a positive net benefit (a view at odds with sustainable transport). It is also an intensely individual perspective; it assumes that society is best rewarded where all individuals can maximise their own utility. In transport, the implications of the capacity limits of infrastructure and the costs of mobility (monetary, environmental and otherwise) are ever present – factors that militate against all individuals having unrestrained mobility. Sustainable transport implies limits on mobility (especially using modes with the highest social and environmental costs) and this necessitates reconsidering how mobility is valued. Community-owned transport will make its greatest contribution not by adding to the total number of journeys taken, but by enabling more trips that are of higher net social value (such as by providing independent mobility to those previously reliant on welfare transport) and by facilitating trips of high-value to replace trips of lower value (such as taking passengers directly to their desired local destination rather than having to travel further and indirectly using public transport). Conviviality, therefore, provides an alternative means for evaluating the value of mobility in a sustainable transport setting.

Conviviality and the language of social connections seem a long way from that usually employed to describe the values of mobility and of public and private transport. Perhaps at first glance, conviviality as a theme of mobility is more than out of place and irrelevant, but illogical to the point of nonsense. A case can be made, however, for considering the place of conviviality in passenger transport. In effect, Illich and Berry are seeking to connect individual choices and acts to their broader social and environmental implications so as to build and reinforce the bonds that bind social groups together as a means to contributing to acting responsibly towards broader society and the environment. Such a perspective is the antithesis of considering every trip a calculation of individual utility that is easily rendered in measurable increments of material value and personal satisfaction. Transport would seem to have a particularly important role in building such conviviality because it can serve as both a source and an enabler of wider social connectivity.

Community-owned transport can contribute to a convivial society in ways that existing public transport cannot and which private transport, for the most part, undermines and degrades. It can do so in the following ways:

- Community-owned transport is based on voluntary community associations which involve social interactions.
- Community-owned transport creates a new social institution for the benefits of its members.
- Community members can articulate and devise strategies and programmes to foster increased social connectivity and environmental protection.
- Such an institution is based in a local community.
- Success in community-owned transport will depend on communication and understanding of members' mobility and other social needs and responding accordingly.

To seek and expect such community-oriented values from community-owned transport is not to argue that community-owned transport needs to be some sort of charity seeking to do good and comprised only of saintly members. Rather, it is that the functions and operations of community-owned transport will contribute to a more convivial society and it is argued here that this is a desirable outcome in and of itself.

Neoliberalism and Urban Mobility

As raised in the Introduction and Chapter 2, neoliberalism is widely embraced by national and state/provincial governments around the world with right wing and many centre-right and centre-left political parties adopting and promoting public policies to give effect to this ideology. Neoliberalism is popular in the English-speaking developed world, notably Australia, Canada, New Zealand, the UK and the US. National economic debates have become less polarised around key issues in these places, with social policy and historical party identity being the major areas of ideological difference. Developing nations have also moved in the neoliberal direction, often promoted by international finance bodies such as the World Bank and IMF, who played a major role in promoting neoliberalism in Latin America, South America and in a number of Asian nations. Where neoliberalism has become well established it is less likely to attract controversy and the practices of neoliberalism have displayed greater longevity than the New Right governments that typically introduced them. Even though some commentators consider the Global Economic Crisis (and several preceding regional economic crises) to be the fault of neoliberal deregulation of the finance and banking sectors, much of the political credibility of neoliberalism has survived intact (Mirowski 2013).

As a political philosophy, as noted earlier, neoliberalism draws strongly on several key themes in liberalism as a way of organising economic markets (Harvey 2005). Markets are at the centre of neoliberal thoughts about the ideal organisation of society; a properly functioning market is essential for the flourishing of all enterprises and maximising individual human welfare. Markets are deemed to be self-organising (through the price mechanism), able to allocate resources according to their best use, thereby achieving the highest economic efficiency. Such markets respond to the preferences of individuals and produce goods and services sought by individuals and firms. Economic productivity is therefore maximised. Markets reward efforts, investments and risk-taking and the resultant distribution of wealth therefore reflects such inputs; inequalities in wealth reflect the differences between individual efforts and talents exercised within markets. Governments are deemed to be inefficient managers of allocation and production in economic systems, as they can never match the capacities of free markets.

Applications of neoliberalism in public policy produced marked effects in governments' approaches to urban services, with the general retreat of scope and scale of government intervention, asset ownership and service provision. Public service has been systematically replaced by private services of various sorts, such as in the provision of community services, education, environmental protection, health, justice and corrections, infrastructure and planning, plus, of course, transport services. Around the world, particularly through the 1980s and 90s, rail and bus public transport services were extensively privatised (particularly in response to rising operating deficits and falling patronage). Typically, privatisation was applied to public transport operations, with the prospect of private financing for public transport infrastructure proving to be far less popular with the private sector (although private financing of road infrastructure has become widespread).

In essence, neoliberal reforms to the public transport sector are characterised by several features. Concern over the indebtedness of public transport systems, over operating fiscal deficits, and the associated levels of state subsidy, particularly in the mass transit modes is a common pre-condition to decision-makers considering major changes to their public transport systems, as described in Chapter 2. This is also often associated with concern over rising costs in the sector and the future expenditure needs for rolling stock replacement, service expansions, dedicated infrastructure, and, to a lesser extent, on-going maintenance costs. Privatisation of public transport and a general exposure of public transport operations to 'market discipline', such as seeking to deregulate labour costs in highly unionised public service providers is one of the most common neoliberal measures taken. Such changes usually mean reducing the role and influence of labour unions, especially where there is history of labour unrest and protests/work stoppages. Deregulation within the transport sector is sought as a way to allow for greater competition in providing services (including labour), such as through private tendering and contracting, with the goal of lowering costs for service provision. That private firms can achieve greater efficiency and improved services than inefficient and poorly operating public sector services is a common belief and rationalisation of neoliberal reform. Around the world, long-standing public transport services were subject to reforms following these principles.

Whether or not this major shift in national politics was the product of an international shift in the political values of the electorates of the developed world is open to question. Accounts of neoliberalism do not associate it with grassroots or civil society movements, but rather as a hegemonic phenomenon advanced by international financial institutions, right-wing think tanks, and conservative political movements around the world (Harvey 2005). What cannot be questioned is that these changes involved major questions of politics and had involved basic aspects of political economy. Central to the politics of these neoliberal reforms to transport, and here we are primarily concerned with public transport, are the reasons put forward by governments to the electorates. These rationalisations are important because, while public transport has often been the subject of political controversy, there was not a public groundswell for neoliberal reforms. Although there is much diversity in the circumstances of cities of developed nations, the 'political script' of neoliberal reform appears to be quite consistent around the world.

Promised benefits of neoliberal reforms usually centred on improved services for travellers, which is hardly surprising. What was sought from neoliberal reforms included (for governments):

- Greater efficiency in key aspects of public transport, including consumer services, planning and management, financial rectitude, performance monitoring and accountability (and greater value for money).
- Lower costs of overall service provision, so that levels of debt would decrease and public subsidies would decrease. In some places, travellers were promised that savings would translate into lower fares.

- Transferring risks to the private sector.
- Increasing or stabilising patronage.
- Stimulating modal shift.
- Fostering innovation.
- Improving the quality of services for consumers in every major aspect and frequently, the promise of an expansion in services.
- Reducing the power of unionised labour.

Consumers were to benefit from: better access to services, improved timetables, improved journey times, new facilities and rolling stock, improved stations and interchanges, responsiveness to demands and greater options and choices.

These changes have not been without controversy. A range of criticisms and concerns have been made (see, for example, Simpson 1996 and Farmer 2011). Typical among the criticisms made of neoliberal reforms include failures to provide improved services and instances of declining services, losses of accountability of service providers and the weakness or absence of institutions to provide public accountability, together with reductions in the information made public about the transport systems because of 'commercial-in-confidence' barriers. Contrary to expectations, there can be an absence of improved returns to governments and a continuation of public subsidies to public transport services. Other complaints relating to financial matters are claims of private operators making windfall profits, the problem 'rent-seeking' by the private operators, and no or limited transfer of finance and investment risks from the public sector to the private sector. There are concerns that private providers seek higher-profit service provision at the expense of the less well-off and reduce services to less profitable routes and locations. Breaking up monopoly transport services can fragment those services and create barriers to integrating transport services. Integrating public transport services within and between modes in systems with greater openness to private sector ownership can require new special bodies and institutional arrangements. Instances of 'regulatory capture' by corporations of governments and public oversight bodies are to the detriment of consumers and the broader public interest. There are also concerns about the ways in which private firms reduce costs after taking over entities from public ownership, notably the loss of employment in transport industries, and lower wages and conditions for employees.

There has been some learning from early failures and mistakes in privatisation and other market reforms in transport. Proponents of neoliberalism have refined the approaches taken to transport and have developed something of a template for privatisation in recognition of many early failures. Many of the unsuccessful schemes simply failed to recognise the character of public transport and devised privatisation schemes that paid insufficient attention to recognising the predatory character of unfettered corporate intervention. Where privatisation appears to have rested is a model with a substantial public role through competitive franchising, regulatory oversight, performance monitoring and enforcement, and protection of consumer, safety and environmental interests.

What does this mean for community-owned transport? There are several implications. First, community-owned transport can serve as an antidote to neo-liberalism by empowering communities who have been on the receiving end of the privatisation of public transport and for whom private transport has proved to be highly inequitable in community terms. Second, for governments wholly entranced by neoliberalism, community-owned transport need not disturb the use and viability of privatised public transport, as both can co-exist. Many of the perceived virtues of privatised public transport at the large scale seem to be identical to those that community-owned transport could exhibit at the local scale.

11 Community-Owned Transport

A Plea to Policy-Makers

To close this volume, this chapter seeks to draw many of the book's themes together around the idea that public policymakers can play an important role in promoting and enabling community-owned transport. For community-owned transport to achieve its full potential and to play a significant role in sustainable transport, the cooperation and direct involvement of governments will be helpful, if not essential. In making the case for changes to public policy to promote community-owned transport, some of the prevailing views in contemporary politics of minimising the role of government are challenged, which in many jurisdictions requires recognising the limits of the neoliberal position. An overriding rationale for such public policy interventions is that many of the barriers facing community ownership are the outcomes of historical and current public policies and established government practices. Overcoming these barriers and reversing some of their effects requires changes to public policy. This chapter finishes with a vision for community-owned transport as sustainable transport.

Recognising the Limits of Neoliberal Approaches to Urban Mobility

In the broad sweep of the modern history of the developed nations, the contest between the right and left constituted the major political narrative, albeit one with many phases and twists and turns. In the English-speaking developed nations at least, neoliberalism has dominated political discourse and mainstream governmental practices since the last quarter of the last century. Some may reject this judgement, pointing to the lasting victories attributed to the left and progressive movements, including feminism, environmentalism, public services in health, education and transport, the continuation of unionism, health and safety protections and other significant measures of social policy and market interventions. All of these accomplishments are indeed victories that can be in large measure attributed to progressive movements and were vigorously opposed by conservative political parties and those with vested interests in maintaining the status quo. In stating that contemporary politics of the OECD nations are those in which the right wing interests are generally ascendant is not to say that the entire landscape reflects their values and policies. However, the governments of the western world, whether of left or right wing origins in recent decades have been broadly neoliberal in economic policy and much of social policy. These governments have not uniformly pursued right wing or conservative policies across all social policy realms and oftentimes have crafted policies within the contours of existing mores around social life and contemporary culture. Commentators have found different interpretations of this phenomenon. For some, it is a maturing of the political landscape and a softening of traditional left/right divisions with coalescence around new accepted norms. For others, the movement of parties with left wing origins into more conservative economic policies is the rise of a 'third way' in politics (Giddens 1998). In these ideological shifts within the established centrist parties in the neoliberal era, the centre-left parties have typically moved to the right, while centre-right parties have generally either also shifted right or stayed relatively in place.

Through the last century or so, urban transport became a routine aspect of governments as the costs of sorting out transport systems and investing in transport infrastructure, particularly major roads, became so high. With this escalation, decision-making over transport became increasingly technocratic and centralised, with its emphasis on meeting the demand for more mobility through increasing the supply of transport infrastructure. Urban transport became increasingly focused on large projects and voters either passively or actively supported programmes of state-guided progress based around high-profile initiatives. In many nations, urban transport continued to attract complaints from voters, particularly from motorists complaining of road congestion and reinforced by powerful lobby groups with interests in road construction, motor vehicle manufacture, transport fuel and motorists' associations. With the rise of sustainable transport, this paradigm has lost its legitimacy, both as a strategy to supply mobility and as a way to govern this activity. Governments around the world are increasingly finding local and urban transport has become a matter of retail politics; there are greater demands for transport options within the sustainable transport portfolio and there is a rejection of governments that decide that transport infrastructure projects and planning are too important to engage the views on voters and constituents.

What are the implications for transport? Neoliberal policies will be the major approach taken by national, state and city governments to transport policy in the developed world in the immediate future and possibly further into the future. Large-scale investment in public transport infrastructure will inevitably occur, especially in heavy rail, but this will be generally minor in relation to the task of shifting mass mobility away from private cars. Public institutions will continue to be regarded as vestiges of big government and the neoliberal imperative of reducing the number of such institutions will remain a doctrine of western governments. Further, governments in developed nations will, in all likelihood, continue to seek ways to reduce public debt and to lower longer-term problems in current accounts. Recurrent public expenditures will therefore continue to be a target for policymakers through the usual strategies of increased pay-for-use of public services and reduced subsidies for public service providers.

In short, these widespread political conditions are generally unfavourable to the cause of sustainable transport. As environmental problems often constitute costs and losses without immediate monetary value, protecting and restoring degraded environments invariably requires the intervention of public and civic action, usually to constrain corporate activities. Most of this burden falls onto governments as shown thorough the process of the 'greening' of the modern state under ecological modernisation (Hajer 1995). This process has been one of accommodating the interests of the corporate sector and those of environmental protection, a process that left undisturbed the basic economic and political structures in industrial society. Against the major gains in environmental protection under ecological modernisation, there are certain environmental issues that have proved particularly intractable, notably those that are not amenable to technological solutions with mature technology alternatives, where the costs of change are high, when problems are complex with multiple causes and where significant social change is required. It should also be noted that neoliberal advocates, politicians and other stakeholders have continually fought against environmental reforms and have, in some instances, had ecological and natural resource protection measures overturned.

Overarching these barriers are the basic concerns that define environmental sustainability as canvassed in an earlier chapter. These barriers are fundamental attributes of industrial capitalist economies that, in summary form, include the commitment to pursue endless economic growth (and associated belief that increasing prosperity resolves environmental problems); the disproportionate allocations of economic, social and environmental costs of economic activity onto those of lower socio-economic status and poorer nations; and the belief in the viability of technological solutions (and available resource substitutions for degraded or exhausted natural resources and ecosystem services). Arguably, these causes of the global environmental crises are the product of global industrial capitalism, rather than of neoliberalism. What makes neoliberalism so antithetical to environmental protection is that it is the foremost and most vigorous source of promotion of this form of capitalism and that, of course, it has become such a widespread political and economic ethos in elected national governments worldwide.

Finding ways of making community-owned transport widely acceptable to policy-makers in neoliberal political settings is a major challenge. Governments will continue to oversee transport systems for the reasons identified earlier (such as preventing market abuses by monopolies and so forth), thereby limiting the extent to which transport systems are exposed to market forces. As discussed earlier, there are aspects of community-owned transport that sustainable transport supporters and civil society may be able to make appealing to elected officials and public sector agencies. For example, community-owned transport can offer many of the claimed benefits of neoliberal approaches of transport without conferring those benefits to corporations or weakening government controls. Responsiveness to customers, flexibility in operations and institutional arrangements, innovation in service provision and perhaps, an operational sensitivity to price and market signals exemplify such benefits. Neoliberal reformers seeking a more passive role for government, with hierarchical control replaced by the use of incentives, flexible arrangements, bargaining and 'learning' strategies could find these features within community-owned transport. Furthermore, community-owned transport can achieve such outcomes in ways that private firms cannot (and especially those that are privatised public agencies), such as they will not require the same degree of government oversight and regulatory control (simply because there is less at stake from the state's perspective, such as protecting public assets and the need to ensure operational performance).

Rejecting many of the principles of governmental control, neoliberal advocates were critical of the regulatory rigidity of state controls, binding legalities of regulations and associated sanctions, and of standardised approaches to policy implementation. Politically, these complaints were allied with political hierarchies, centralised authority and the use of formal institutions to exert control over policy realms. Their reforms sought to break down these features through strategies such as using flexible approaches to policy implementation, delegating state functions to corporations and having agreements and arrangements between states and corporations. Arguing for a retreat and softening of state authority by neoliberals paradoxically opens the potential for non-government organisations, community groups and other non-state and non-corporate entities.

Given the aforementioned limits of markets in urban transport, the greater use of market-based approaches is quite limited and proscribed. Logically, meeting the goals of conservative reformers of urban transport, such as having greater responsiveness to the wishes of individuals, more dispersed authority, more flexible implementation and less-centralised decision-making leads to the greater use of community-based institutions. These same attributes can be achieved through private carsharing and ridesharing business, but with community-owned transport the benefits accruing to the operators are returned to community organisations, rather than to corporations and independent contractors. Further, as suggested earlier, community-owned transport can operate in niche markets without disturbing the status quo of those serving mass markets in public transport.

Promoting Community Ownership of Transport through Public Policy

Ideally, community-based initiatives, such as community ownership of transport, springs forth spontaneously without interference from, or promotion by, governments. Historically, the effects of path dependence resist major reform in urban transport and community-owned transport will have to overcome many barriers in order to become more widespread. Institutional inertia, laws and regulations supporting existing arrangements, political influences of vested and entrenched interests, established preferences for technical and economic solutions and ingrained scepticism of initiatives from non-professional groups by professionals – as canvassed in previous chapters – constitute some of these barriers. Without assistance, communities will have a difficult time succeeding against these factors, most of which were created or enhanced through public policy, either directly or indirectly. Reforming public policy will greatly assist the cause of community ownership, if only to reduce the existing prejudices of prevailing decision-makers.

We have an abundance of models for community ownership, not only for transport but also for a wide range of services, as described. These cover a wide array of types of social organisations, motivations and programmes of activity. Given that these include social enterprises, cooperatives and charitable organisations, it is clear that further growth in community-owned transport will only see further diversification in how such ownership is organised. Whilst some of these services are independent of public transport systems, there are many exemplars of integration between local community transport and larger public transport systems. Further, in the ecovillage movement, there are many exemplars of the organisation of social life according to principles of environmentally sustainable living. Critically, ecovillages also model arrangements for collective decision-making applicable for community-owned transport.

Government powers in the three main spheres (national, state or province and local) and the inter-governmental sphere (such as whole-of-city government) cover a great range of activities and responsibilities, with potentially many of these having some impact on community-owned transport. Transport public policy impinges directly on these interests, such as the responsibilities over passenger services, licences for operating vehicles, vehicle registration, vehicle standards and transport infrastructure. As has been shown earlier in this volume, there are also many other aspects of public policy that could assist the cause of community-owned transport, including responsibilities over business, civic infrastructure, community services, employment and environment. Because of the complexity of this problem, it is best to identify here several broad measures that could be taken by a variety of state agencies and spheres of government. Specifically, public policy reform to assist community ownership could take several forms, as raised in some of the preceding chapters, such as in the following areas.

Integrating Community-Owned Transport and Public Transport

Much of current community-based transport is organised as demand-responsive transport, especially in rural areas, as described above. Whilst there are concerns that demand-responsive transport works against the organisation of integrated transport, it is clear from examples in Switzerland and elsewhere that these two models of service provision can be made compatible. Facilitating easy transfer between community-owned transport and public transport and active transport nodes requires interchanges at key locations, especially those that provide ready access to the wider public transport network of the city and wider region. Part of the success of these interchanges depends on appropriate design, and features such direct and easy transfers between services, allowance for pedestrian movements that are separate from vehicles, good connections between the interchange facilities and surrounding buildings and services, appropriate ancillary services within the interchange facility, effective information and advisory systems and bicycle parking. Wider integration involves creating governance institutions to ensure on-going management and coordination between different organisations, shared information systems providing a common service for all travellers and coordination of transport services to enable easy and efficient transfers between different services. Integration is often not welcomed by commercial operators, and sometimes also by government agencies, so leadership and strategic direction is needed by suitable government agencies with the resources and authority to ensure compliance and eventual success. Critically, the element of success in integration planning is the role of the government in ensuring that it is undertaken.

Planning for Land Use and Transport

Planning can reduce the need for travelling, discourage car use and encourage active transport through strategic policy, structure planning and land use planning, development codes and project approval processes by various measures. In principle, these measures can make public transport and active transport more convenient for users by placing transport services closer to where they are needed, providing for a clustering of transport destinations and facilitating direct movement between them. Ways to achieve these goals include linking land uses and transport services, concentrating various land uses in central locations, having a mix of land uses in centres, ensuring connectivity between streets, ensuring streets have active transport facilities (i.e., footpaths and bicycle lanes/paths) and road management to promote public and active transport.

Recognising Special Needs

Community-owned transport can be shown to operate under all sorts of regulatory regimes and government settings; however, all transport operators face a considerable array of rules and regulations. Authorities and community bodies could work together to establish appropriate regulatory settings to meet the needs of community-owned transport organisations, covering such matters as insurance, accreditation and registration. Governments are having to respond to commercial collaborative consumption services and community-owned transport could be made part of a wider regulatory response.

Supporting Community Organisations

Governments could make available resources for community ownership groups to assist in organising and managing transport operations. Public and private transport rarely pays its way; public transport is invariably subsidised by public funds and private transport benefits from public road infrastructure, governance arrangements and other direct and indirect public subsidy. Subsidising community groups to provide transport services is far from an affront to established practices; rather, subsidy has been a large part of the economics of urban transport for the entire modern era. From a historical perspective, decisions not to subsidise community-owned transport would prove anomalous in the transport industry;

a case for public subsidy of community-owned transport, even in this neoliberal era, is strong.

An Overview of the Benefits of Community-Owned Transport

Community-owned transport can contribute to sustainable transport, offering a number of benefits. It can play a part in the overall quest for shifting from models of mobility based on private motor vehicles using fossil fuels (automobility) to models of sustainable transport, featuring modest levels of mobility, high levels of public transport and active transport and urban land uses integrated with transport infrastructure. Many themes of sustainable transport could be addressed by community ownership, as it offers a means to assist this transition by providing a means to provide access to shared transport, public transport and bicycles. By improving access to more sustainable transport modes, there are opportunities to increase the use of such services and to replace trips by private motor vehicles.

Additional and improved access to mobility that is currently out of reach through community-owned transport improves overall mobility within and between cities. For those with poor access to mobility, community-owned transport could go some way to reducing mobility inequality, especially for groups without car access, including those of lower socioeconomic status, women, the elderly, children and teenagers. For those afflicted with car-dependency, community-owned transport could provide alternative mobility without having to rely on public transport (which is typically poor where there is high car-dependency).

Economically, community-owned transport has attractive features. Using existing transport infrastructure, community-owned transport increases the use and value of such assets. Rather than relying on investments in transport infrastructure to increase mobility, community-owned transport offers a means to increase mobility at lower costs using existing infrastructure. Where community-owned transport substitutes multiple passenger vehicles for single-occupant vehicles, the resulting gain will bring economic benefits to transport systems and, most likely, to those travelling. A wider set of mobility choices and alternatives for consumer spending would be available for community-owned transport users. While large public transport systems have to deal with the problems of the poor economies of scale in adding extra services in locations of relatively low and dispersed demand, the economics of community-owned transport are not necessarily confined in the same way. Indeed, the overhead costs encountered by public transport services are not always the same imposition on community-owned transport that, for example, may have no need for broad-scale advertising and marketing, elaborate ticketing systems and inspections, data collection and analysis, system planning and corporate overheads. Rather than being costly because they are small scaled, community-owned transport can achieve economic benefits in ways that elude large-scale systems, such as serving niche markets and providing integrated mobility services.

Socially, community-owned transport offers an alternative to relying on public transport to provide mobility for those not using private transport. Because community-owned transport can provide services based on the local mobility needs of communities, the users are not subject to a uniform model of service as determined by government institutions or corporations. Community-owned transport can provide greater flexibility in meeting local demands and offering greater autonomy for the community owners than is possible under public transport. There may also be other benefits to communities over and above additional mobility arising from the conviviality of communal ownership. Large public transport services are remote, bureaucratic and regimented; community-owned transport is local, accountable, flexible, needs-driven and empowers community organisation. Members of community-owned transport organisations can express their views, preferences and needs in ways that directly shape the policies, management and operations of community-owned transport services thereby offering models of collaborative and participatory engagement in transport service institutions unavailable in any other form of ownership.

Environmentally, community-owned transport has a number of benefits that suit the goals of sustainable transport. Directly, community-owned transport can provide for mobility in multiple passenger vehicles and provide an alternative to private car use, it can use vehicles with lower environmental impacts and it can increase the use and viability of public transport and active transport through trip integration strategies. By providing local accessibility, the need for higher speed and more energy expensive trips to reach further destinations is reduced; local transport can be slower and more energy efficient without reducing service quality. There are also indirect and more nuanced ways in which it can support environmental goals. Relying on a narrow set of indicators to assess transport system performance has given way to a wider set of criteria, including a range of social, cultural and environmental values – a trend that supports community-owned transport services. Community-owned transport also promotes local services and businesses and greater local accessibility can bolster local economies and the viability of local businesses and commercial activities. Overall, this effect can help reduce the overall need for travel and reduce travel distances compared to the longer trips to access widely dispersed destinations. Local cultural activities can also be bolstered through improved local mobility and these can be part of wider efforts to pursue environmental sustainability programmes.

As the collaborative consumption industry has shown, where there are failings in conventional markets and with sufficient consumer demand, there are opportunities for new businesses. Oftentimes, these new businesses are not entirely original or novel, but have an advantage over existing businesses and business models that enables them to the 'disrupt' the existing market. Ridesharing companies, for instance, are a merging of conventional taxi businesses and informal transport that uses contemporary communications and information technologies as the enabler of a new business model for mobility. Urban transport is a market rife with potential new opportunities for passenger mobility. Latent demand in urban transport is high, as argued earlier, because of the diversity, fluidity and spontaneity of the need to travel and the inability of transport systems to ever satisfy all such demands. If community ownership does not develop in the coming decades, it is highly likely that there will be a host of new disruptive businesses emerging to find profit by identifying new niche markets that cannot be satisfied by the mass mobility provided by public transport and privately-owned motor vehicles. Losing such opportunities will be to the detriment of communities in general, for the new firms providing mobility that might have been provided by community-owned entities will not furnish the net community benefits available from community ownership.

Community-owned transport also offers a partial solution to the problem of informal transport. Although informal transport is the transport of last resort where private vehicle ownership is too expensive and public transport is inadequate, it is usually accepted as a necessity in impoverished cities and is generally illegal in developed nations. A viable alternative to informal transport is community-owned transport and it could assist in providing a form of low-cost and flexible mobility in developing nations, but also offers the advantages of conforming to safety regulations, as it does not make operational decisions solely based on profit maximisation and does not place a significant demand on government resources.

Joining a community-owned transport group is unlikely to be motivated by any formal political objectives, but community ownership does have political significance and can be viewed from a political perspective. Although policy discussions over urban transport strategy and practices are often dominated by neoliberal thinking, community-owned transport offers collective solutions as an alternative to individual transport and monopolised state and private public transport. For supporters of the new economy ideas, community-owned transport offers some of the attractive features of the collaborative economy and also provides a barrier against some of its downsides (such as the slide from a formal regulated transport industry into an unregulated informal transport sector). On the positive side, community-owned transport offers an alternative to owning motor vehicles, can use information and communication technologies to inform users and arrange services and offers flexibility and the possibility of innovation as a result of the close relationships made possible in collaborative arrangements. Community-owned transport can also be fitted into a state regulatory system and meet safety, operating and other standards required of transport services.

A Vision for Community-Owned Transport

Some time in the future, the idea of community ownership of transport will have become so common around the world as to be completely unremarkable. There will be many versions and approaches to community ownership, operating at many different scales and using many different modes of transport. Community-owned services will integrate seamlessly with the usual public transport services and link local transport services with wider urban transport networks, so that local users will have access across cities and regions. Community-owned transport organisations would be viewed as places of innovation and creativity in providing urban mobility and other services.

Community-owned transport users would comprise a broad spectrum of society; it will provide services without stigma and appeal to members of all socioeconomic groups. Some of those who were formerly restricted to welfare and charitable transport services will now have access to affordable local mobility services and have far greater levels of service available to them. Households that previously depended on car ownership for mobility will have choices and opportunities for travel without using a private motor vehicle.

Environmentally sustainable transport goals will be greatly assisted by the widespread use of community-owned transport that promotes greater use of public and active transport, reduces private car use and offers services that have lower environmental impacts than the private car. Land use planning will have responded to the opportunities provided by community-owned transport as part of the wider efforts to plan transport and land use simultaneously around sustainable urban living objectives.

Membership and involvement in community-owned transport will have become commonplace and routine in the same way that other community-based activities have become, with members actively involved in the planning, managing and operating of local transport services in accordance with their own preferences, needs and values. Autonomy and self-regulation will be central features of these organisations, with accountability to their membership and the wider community. Entrepreneurs will have found ways to build new businesses around the opportunities provided by community-owned transport, creating local employment and services to communities that multiply the benefits created by local transport systems. Community-owned transport is not considered as a charity or form of welfare, but as an enterprise whose members find ways to contribute to an array of community and individual needs.

In Closing

Community-owned transport is not a panacea for the economic, social and environmental failings of urban transport, but neither is it without worth or impossible to achieve. For too long, mass urban transport has been a tug-of-war between public and private transport, between private firms and state corporations, and between individual needs and mass services. Much of this contestation has failed to produce sustainable transport. Community-owned transport can play a role in urban mobility and perhaps a major role – who knows? One thing is clear – transport is proving largely intractable in shifting to more sustainable practices. There are numerous barriers to change and in some ways the problem is a chain of linked problems; the resolution of one or more links alone is insufficient to produce systemic change. Amongst these linked problems are the energy sources (fossil fuels) and the internal combustion engine, the type of mobility offered by individual motorised mobility, the necessity of mobility within modern cities dictated by economic geography and land use planning, public transport and the challenges of dispersed and episodic demand, urban population growth and the specialised infrastructure needs of active transport. To these can be added the economic and institutional problems, such as institutional inertia and path dependency, corporate rent seeking, failures of public accountability, bureaucratic complacency and satisficing strategies. With every turn of the wheel, new solutions to urban transport arise. Hydrogen as a green energy source, the electrification of mobility, increased privatisation of public transport and promoting higher-rise city centres have all had their moment in the sun as solutions to unsustainable transport. None of these solutions have worked to transform urban transport, for all sorts of reasons, but their failures show us that changing a few components will not transform a complex machine.

Community-owned transport is a different sort of solution to the problem of mobility and can play a role in promoting sustainable transport. There have been numerous innovations in many aspects of transport, particularly in vehicle and energy systems, in the financing, the governing of public transport under privatisation strategies and in the new businesses based on sharing economy strategies. But there has been relatively little attention given to innovations in ownership and the examples of community ownership of transport. Within the broad concept of community ownership are many of the features that can liberate urban transport from its stolid insistence on traditional models of ownership by creating urban transport systems that are more diverse in character, more adaptive and nimble, more innovative and creative and more attuned to the goals of a mobility that is more socially equitable and less ecologically harmful.

So, why not community-owned transport? Not all environmental initiatives have to be managed from 'above'. We do not need to resolve the major issues facing global ecology in order to organise our communities and ourselves into acting in ways that do less harm to the environment, both globally and locally. Ultimately, the move to community ownership offers a small emancipation. We should be allowed and encouraged to strengthen our communities in practical ways to provide local mobility. In doing so, we can be liberated and empowered in making choices about mobility within and beyond established transport networks. Such liberation is not just about enhancing personal mobility, but enhancing community mobility; it is not just about consuming mobility but producing mobility in environmentally responsible ways; it is not just about increasing the personal mobility of the mobility-rich, but enhancing mobility to those suffering mobility disadvantage and it is not only about greater autonomy but also about greater responsibility. Community-owned transport entails changing our thinking about urban mobility from being about individual demands for consumption to contributing to, and producing, community mobility. It is a change, therefore, from a mobility of taking and consuming to a mobility of connecting, learning, growing and giving. Mobility is a form of commons, after all.

References

Aleklett, K., 2012. *Peeking at peak oil*. New York, NY: Springer.
Aleklett, K., Höök, M., Jakobsson, K., Lardelli, M., Snowden, S. and Söderbergh, B., 2010. The age of peak oil: Analysing the world oil production reference scenario in World Energy Outlook 2008, *Energy Policy*, Vol. 38 (3): 1398–1414.
Alexander, D., 1990. Bioregionalism: Science or sensibility? *Environmental Ethics*, Vol. 12 (2): 161–173.
Ameratunga, S., Hijar, M. and Norton, R., 2006. Road-traffic injuries: Confronting disparities to address a global-health problem, *Lancet*, Vol. 367 (9521): 1533–1540.
Anable, J., 2005. 'Complacent car addicts' or 'aspiring environmentalists'? Identifying travel behaviour segments using attitude theory, *Transport Policy*, Vol. 12 (1): 65–78.
Arnstein, S.R., 1969. A ladder of participation, *Journal of the American Planning Association*, Vol. 35 (4): 216–224.
Arthur, W.B., 1994. *Increasing returns and path dependence in the economy*. Ann Arbor, MI: University of Michigan Press.
Auld, D.A.L., 2013. *Accountability denied: The global biofuel blunder*. Victoria, BC: Friesen Press.
Bahro, R., 1986. *Building the green movement*. London: Heretic/GMP.
Banister, D., 1994. *Transport planning*. London: E and FN Spon.
Banister, D., 1998. *Transport planning and the environment*. London: Routledge.
Banister, D., 2005. *Unsustainable transport: City transport in the new century*. New York, NY: Routledge.
Barton, H., 2005. Conflicting perceptions of neighbourhood. In: H. Barton, ed., *Sustainable communities: The potential for eco neighbourhoods*. London: Earthscan, 3–18.
Baxter, B., 1999. *Ecologism: An introduction*. Washington, DC: Georgetown University Press.
Beatley, T., 1999. *Green urbanism: Learning from European cities*. Washington, DC: Island Press.
Beaumont, M. and Freeman, M., eds., 2007a. *The railway and modernity: Time, space, and the machine ensemble*. Oxford, UK: Peter Lang.
Beaumont, M. and Freeman, M., 2007b. Introduction: Tracks to modernity. In: M. Beaumont and M. Freeman, eds., *The railway and modernity: Time, space, and the machine ensemble*. Oxford, UK: Peter Lang, 13–43.
Benton, T., ed., 1996. *The greening of Marxism*. New York, NY: Guildford.
Berry, W., 2008. Faustian economics: Hell hath no limits, *Harper's Magazine*, 18 May: 1–7.
BITRE, Bureau of Infrastructure, Transport and Regional Economics, 2014. *Long-term trends in urban public transport*. Information sheet 60. Canberra: BITRE.

Böhm, S., Jones, C., Land, C. and Patterson, M., eds., 2006. *Against automobility*. Oxford: Blackwell.

Bollier, D., 2013. *Silent theft: The private plunder of our common wealth*. London: Routledge.

Bonsall, P., Jopson, A., Pridmore, A., Ryan, A. and Firmin, P., 2002. *Car share and car clubs: Potential impacts*. Final report to the DTLR and the Motorists Forum, London.

Bookchin, M., 1980. *Toward an ecological society*. Montreal: Black Rose Books.

Bookchin, M., 1986. *The modern crisis*. Philadelphia, PA: New Society Publishers.

Boschmann, E.E. and Kwan, M.-P., 2008. Toward socially sustainable urban transportation: Progress and potentials, *International Journal of Sustainable Transportation*, Vol. 2:138–157.

Botsman, R. and Rogers, R., 2010. *What's mine is yours: The rise of collaborative consumption*. New York, NY: HarperCollins.

BP., 2015. *BP energy outlook 2035*. Available at: www.bp.com/content/dam/bp/pdf/energy-economics/energy-outlook-2015/bp-energy-outlook-2035-booklet.pdf

Brown-West, O.G., 2007. *In defense of the big dig: How politics affected the planning, design and construction of the Boston Central Artery*. Marshfield, MA: Diabono Consolidated Inc.

BTS, Bureau of Transportation Statistics, 2013. *Transportation statistics annual report 2012*. Washington, DC: U.S. Department of Transportation.

Buck, S., 1998. *The global commons: An introduction*. Covelo, CA: Island Press.

Byrne, J. and Glover, L., 2002. A common future or towards a future commons: Globalization and sustainable development since UNCED, *International Review for Environmental Strategies*, Vol. 3 (1): 5–25.

Byrne, J., Martinez, C. and Ruggero, C., 2009. Relocating energy in the social commons: Ideas for a sustainable energy utility. *Bulletin of Science, Technology and Society*, Vol. 29 (2): 81–94.

Cahill, M., 2010. *Transport, environment and society*. London: McGraw-Hill International.

Cairns, S., 2011. *Assessing cars: Different ownership and use choices*. London: Royal Automobile Club Foundation.

Cairns, S., Sloman, L., Newson, C., Anable, J., Kirkbride, A. and Goodwin, P., 2004. *Smarter choices: Changing the way we travel*. Report for the UK Department for Transport.

Calthorpe, P., 1993. *The next American metropolis: Ecology, community, and the American dream*. New York, NY: Princeton Architectural Press.

Caprotti, F., 2015. *Eco-cities and the transition to low carbon economies*. Basingstoke: Palgrave Macmillan.

Carlson, D., 1995. *At roads end: Transportation and land use choices for communities*. Washington, DC: Island Press.

Carplus, 2015. *Community car clubs handbook: A practical guide for individuals and community groups*.

Carr, S., Francis, M., Rivlin, L.G. and Stone, A.M., 1992. *Public space*. Cambridge, UK: Cambridge University Press.

Carter, I., 2001. *Railways and culture in Britain: The epitome of modernity*. Manchester: Manchester University Press.

Carter, N., 2001. *The politics of the environment: Ideas, activism, policy*. Cambridge, UK: Cambridge University Press.

Castells, M., 1996. *The rise of network society*. Malden, MA: Blackwell.

Caufield, B., 2009. Estimating the environmental benefits of ride-sharing: A case study of Dublin, *Transportation Research Part D*, Vol. 14 (7): 527–531.
Cervero, R., 1998. *The transit metropolis: A global inquiry*. Washington, DC: Island Press.
Cervero, R., 2000. *Informal transport in the developing world*. Nairobi: UN Centre for Human Settlements (Habitat).
Cervero, R. and Golub, A., 2007. Informal transport: A global perspective, *Transport Policy*, Vol. 14 (6): 445–457.
Chapman, L., 2007. Transport and climate change: A review, *Journal of Transport Geography*, Vol. 15: 354–367.
Cheape, C.W., 1980. *Moving the masses: Urban public transit in New York, Boston, and Philadelphia 1880–1912*. Boston, MA: Harvard University Press.
Cherry, C. and Cervero, R., 2005. Use characteristics and mode choice behaviour of electric bike users in China, *Transport Policy*, Vol. 14 (3): 247–257.
Chiras, D. and Wann, D., 2003. *Superbia: 31 ways to create sustainable neighborhoods*. Gabriola Island: New Society Publishers.
City Carshare, (no date). City carshare background. Available at: http://citycarshare.org/wp-content/uploads/2014/10/City-CarShare-Press-Kit-10-19-14.pdf
CIT, Commission for Integrated Transport, 2008. *A new approach to rural public transport*. CIT: London.
Clarke, D., 2007. *The battle for barrels: Peak oil myths and world oil futures*. London: Profile Books.
Coffin, A.W., 2007. From roadkill to road ecology: A review of the ecological effects of roads, *Journal of Transport Geography*, Vol. 15 (5): 396–406.
Cox, M.G., Arnold, G. and Villamayer Tomás, S., 2010. A review of design principles for community-based natural resource management, *Ecology and Society*, Vol. 15 (4): 38. Available at: www.ecologyandsociety.org/vol15/iss4/art38/
Curtis, F., 2009. Peak globalization: Climate change, oil depletion and global trade, *Ecological Economics*, Vol. 69 (2): 427–434.
Daly, H., 1977. *Steady-state economics*. San Francisco, CA: W.H. Freeman.
Daly, H., 1990. Sustainable growth: An impossibility theorem, *Development*, No. 3–4: 45–47.
David, P., 1985. Clio and the economics of QWERTY, *American Economic Review*, Vol. 75 (2): 332–337.
Davis, S.C., Diegel, S.W. and Boundy, R.G., 2014. *Transport energy data book: Edition 33*. Oak Ridge, TN: US Department of Energy.
Davison, L., Enoch, M., Ryley, T., Quddus, M. and Wang, C., 2014. A survey of demand responsive transport in Great Britain, *Transport Policy*, Vol. 31: 47–54.
Deffeyes, K.S., 2009. *Hubbert's peak: The impending world oil shortage*. Princeton, NJ: Princeton University Press.
DeMaio, P., 2009. Bike-sharing: History, impacts, models of provision, and future, *Journal of Public Transportation*, Vol. 12 (4): 41–56.
Dennis, K. and Urry, J., 2009. *After the car*. Cambridge, UK: Polity.
Dimitirou, H.T., ed., 1990. *Transport planning for third world cities*. London: Routledge.
Dimitirou, H.T. and Gakenheimer, R., eds., 2011. *Urban transport in the developing world: A handbook of policy and practice*. Cheltenham, UK: Edward Elgar.
Dobson, A., 2000. *Green political thought*. Third edition. London: Routledge.
Docherty, I., Shaw, J. and Gather, M., 2004. State intervention in contemporary transport, *Journal of Transport Geography*, Vol. 12 (4): 257–264.

DOT, Department of Transport, 2014. Transport and accessibility to services in rural areas: January 2014. Available at: www.gov.uk/government/statistics/transport-and-travel-in-rural-and-urban-areas

Dryzek, J.S., 1997. *The politics of the earth: Environmental discourses*. Oxford, UK: Oxford University Press.

Duany, A. and Plater-Zyberk, E. with Speck, J., 2001. *Suburban nation: The rise of sprawl and the decline of the American dream*. New York, NY: Farr, Straus and Giroux.

Dudley, G. and Richardson, J., 2004. *Why does policy change? Lessons from British transport policy 1945–99*. London: Routledge.

Dunn, J.A., 1998. *Driving forces: The automobile, its enemies, and the politics of mobility*. Washington, DC: Brookings Institution Press.

Dyer, G., 2008. *Climate wars: The fight for survival as the world overheats*. Toronto: Random House Canada.

Eckersley, R., 1992. *Environmentalism and political theory: Towards an ecocentric approach*. London: University College London Press.

Elgowainy, A., Burnham, A., Wang, M., Molburg, J. and Rousseau, A., 2009. *Well-to-wheels energy use and greenhouse gas emissions analysis of plug-in hybrid electric vehicles*. Oak Ridge, TN: Argonne National Laboratory, US Dept. of Energy.

Ellis, E. and McCollom, B., 2009. *Guidebook for rural demand-response transportation: Measuring, assessing, and improving performance*. TCRP report 136. Washington, DC: Transportation Research Board.

Enoch, M., Potter, S., Parkhurst, G. and Smith, M., 2004. *INTERMODE: Innovations in demand responsive transport*. Prepared for Department for Transport and the Greater Manchester Passenger Transport Executive. Department of Transport, London.

Enoch, M., Potter, S., Parkhurst, G. and Smith, M., 2006. Why do demand responsive transport systems fail? *Transportation Research Board 85th Annual Meeting*, 22–26 January, Washington DC.

Enoch, M.P., 2002. Car clubs: Lessons from the Netherlands and San Francisco, *Traffic Engineering and Control*, Vol. 43 (4): 131–134.

Estache, A. and de Rus, G., 2000. The regulation of transport infrastructure and services: A conceptual overview. In: A. Estache, ed., *Privatization and regulation of transport infrastructure: Guidelines for policymakers and regulators*. Washington, DC: World Bank, 5–50.

European Commission, 2009. *A sustainable future for transport: Towards an integrated, technology-led and user-friendly system*. Luxembourg: Publications Office of the European Union.

European Commission, 2011. *A roadmap for moving to a competitive low carbon economy in 2050*. Communication from the Commission to the European Parliament, the Council, the European Economic and Social Committee and the Committee of the Regions. Brussels. Available at: http://eur-lex.europa.eu/legal-content/en/TXT/?uri=CELEX%3A52011DC0112

ECMT, European Conference of Transport Ministers, 2003. *Implementing sustainable urban travel policies: National reviews*. Paris: OECD.

ECMT, European Conference of Transport Ministers, 2007. *Cutting transport CO_2 emissions: What progress?* Paris: OECD.

EEA, European Environment Agency, 2013. *Air quality in Europe: 2013 report, EEA report No. 9/2013*. Luxembourg: Publications Office of the European Union.

Farmer, S., 2011. Uneven public transportation development in neoliberalizing Chicago, USA, *Environment and Planning A*, Vol. 43 (5): 1154–1172.

Farmer, S., 2013. On the move: The neoliberalization of US public transportation, *Harvard International Review*, Vol. 35 (2): 60–64.
Farrington, J. and Farrington, C., 2005. Rural accessibility, social inclusion and social justice: Towards conceptualisation. *Journal of Transport Geography*, Vol. 13 (1): 1–12.
Featherstone, M., Thrift, N. and Urry, J., eds., 2005. *Automobilities*. London: Sage.
Felson, M. and Spaeth, J.L., 1978. Community structure and collaborative consumption, *American Behavioral Scientist*, Vol. 21 (4): 614–624.
Fishman, R., 1987. *Bourgeois utopias: The rise and fall of suburbia*. New York, NY: Basic Books.
Fishman, L. and Wabe, J.S., 1969. Restructuring the form of car ownership: A proposed solution to the problem of the motor car in the United Kingdom, *Transportation Research*, Vol. 3 (4): 429–442.
Fishman, E., Washington, S. and Haworth, N., 2013. Bike share: A synthesis of the literature, *Transport Reviews*, Vol. 33 (2): 148–165.
Flink, J.J., 1988. *The automobile age*. Cambridge, MA: MIT Press.
Freeman, M., 1999. *Railways and the Victorian imagination*. New Haven, CT: Yale University Press.
Frischmann, B.M., 2005. Infrastructure commons, *Michigan State Law Review*, Vol. 89 (Spring): 121–136.
Frischmann, B.M., 2012. *Infrastructure: The social value of shared resources*. Oxford: Oxford University Press.
Furness, Z., 2010. *One less car: Bicycling and the politics of automobility*. Philadelphia, PA: Temple University Press.
Gansky, L., 2010. *The mesh: Why the future of business is sharing*. New York, NY: Penguin.
Garrison, W.L. and Levinson, D.M., 2014. *The transportation experience: Policy, planning, and deployment*. Oxford: Oxford University Press.
Geerlings, H., Shiftan, Y. and Stead, D., 2012. The complex challenge of transitions towards sustainable mobility: An introduction. In: H. Geerlings, Y. Shiftan and D. Stead, eds., *Transition towards sustainable mobility: The role of instruments, individuals and institutions*. Aldershot, UK: Gower Publishing, 1–9.
Giaoutzi, M. and Nijkamp, P., eds., 2008. *Network strategies in Europe: Developing the future for transport and ICT*. Aldershot: Ashgate.
Giddens, A., 1990. *The consequences of modernity*. Stanford, CA: Stanford University Press.
Giddens, A., 1998. *The third way: The renewal of social democracy*. Cambridge, UK: Polity Press.
Gilbert, R. and Perl, A., 2010. *Transport revolutions: Moving people and freight without oil*. Second Edition. Gabriola Island, BC: New Society Publishers.
Givoni, M. and Banister, D., 2010. *Integrated transport: From policy to practice*. Abingdon, Oxon: Routledge.
Glover, L., 2011. Public transport as a common pool resource, *Australasian Transport Research Forum 2011*, Adelaide, Australia, 28–30 September.
Glover, L., 2012. Public policy options for the problem of public transport as a common pool resource, *Australasian Transport Research Forum 2012*, Adelaide, Australia, 26–28 September.
Glover, L., 2013. A communal turn for transport? Integrating community-owned transport and public transport for sustainable transport, *Proceedings of the People and the Planet Conference*, RMIT University, Melbourne, 2–4 July.

Goldman, T. and Gorman, R., 2006. Sustainable urban transport: Four innovative directions, *Technology in Society*, Vol. 28 (1): 261–273.

Goodman, A., Green, J. and Woodcock, J., 2014. The role of bicycle sharing systems in normalising the image of cycling: An observational study of London cyclists, *Journal of Transport and Health*, Vol. 1 (1): 5–8.

Goodwin, P., 2012. *Peak travel, peak car and the future of mobility: Evidence, unresolved issues, policy implications, and a research agenda.* Discussion paper 2012/13. International Transport Forum.

Gössling, S. and Cohen, S., 2014. Why sustainable transport policies will fail: EU climate policy in the light of transport taboos, *Journal of Transport Geography*, Vol. 39: 197–207.

Graham, S. and Marvin, S., 2001. *Splintering urbanism: Networked infrastructures, technological mobilities, and the urban condition.* London: Routledge.

Granberg, M. and Åström, J., 2010. Civic participation and interactive decision-making: A case study. In: E. Amnå, ed., *Normative implications of new forms of participation from democratic policy processes*. Berlin: Nomas Verlag: 53–65.

Grant, J.L., 2009. Theory and practice in planning the suburbs: Challenges to implementing new urbanism, smart growth, and sustainability principles, *Planning Theory and Practice*, Vol. 10 (1): 11–13.

Gray, D., Farrington, J.H. and Kagermeier, A., 2008. Geographies of rural transport, In: R. Knowles, J., Shaw and I. Docherty, eds., *Transport geographies: Mobilities, flows, and spaces*. Malden, MA: Blackwell, 102–119.

Grübler, A., 2003. *Technology and global change.* Second Edition. Cambridge, UK: Cambridge University Press.

Guha, R., 2000. *Environmentalism: A global history*. New York, NY: Longman.

Hajer, M.A., 1995. *The politics of environmental discourse: Ecological modernization and the policy process*. Oxford: Oxford University Press.

Hall, P., 1980. *Great planning disasters*. Berkeley, CA: University of California Press.

Hall, P., 1988. *Cities of tomorrow: An intellectual history of urban planning and design in the twentieth century*. Oxford: Blackwell.

Hansen, J., 2009. *Storms of my grandchildren: The truth about the coming climate catastrophe and our last chance to save humanity*. New York, NY: Bloomsbury.

Hardin, G., 1968. The tragedy of the commons, *Science*, Vol. 162: 1243–48.

Harvey, D., 1989. *The condition of postmodernity: An inquiry into the origins of cultural change*. Cambridge, MA: Blackwell.

Harvey, D., 2005. *A brief history of neoliberalism*. Oxford: Oxford University Press.

Hess, C. and Ostrom, E., 2003. Ideas, artifacts, and facilities: Information as a common-pool resource, *Law and Contemporary Problems,* Vol. 66 (1 & 2): 111–145.

Hill, J., Nelson, E., Tilman, D., Polasky, S. and Tiffany, 2006. Environmental, economic, and energetic costs and benefits of biodiesel and ethanol biofuels, *Proceedings of the National Academy of Science*, Vol. 103 (30): 11206–11210.

Hillary, G., 1955. Definitions of community: Areas of agreement, *Rural Sociology*, Vol. 20 (2): 111–123.

Hood, C., 1991. A public management for all seasons? *Public Administration*, Vol. 69: 3–19.

Hoogma, R., Kemp, R., Schot, J. and Truffer, B., 2005. *Experimenting for sustainable transport: The approach of strategic niche management*. London: Routledge.

Hook, W. and Fabian, B., 2009. *Regulation and design of motorized and non-motorized two-and-three-wheelers in urban traffic*. New York, NY: Institute for Transportation and Development Policy.

Horowitz, A., 2014. Uber gets failing grade from Better Business Bureau, *Huffington Post*, 11 October 2014.

Hothersall, D.C. and Salter, R.J., 1977. *Transport and the environment*. London: Crosby Lockwood Staples.

IEA, International Energy Agency, 2005. *Saving oil in a hurry*. Paris: OECD.

IEA, International Energy Agency, 2006. *Energy technology perspectives: Scenarios and strategies to 2050*. Paris: IEA.

IEA, International Energy Agency, 2012. *Energy technology perspectives 2012: Pathways to a clean energy system*. Paris: IEA.

IEA, International Energy Agency, 2013. *World energy outlook 2013*. Paris: IEA.

IEA, International Energy Agency, 2014. *CO_2 emissions from fuel combustion: Highlights*. Paris: IEA.

IEA, International Energy Agency, 2015a. *Energy and climate change*. Paris: IEA.

IEA, International Energy Agency, 2015b. *Tracking clean energy progress 2015*. Paris: IEA.

Illich, I., 1974. *Energy and equity*. London: Calder and Boyars.

IPCC, Intergovernmental Panel on Climate Change, 2014a. *Climate change 2014: Mitigation of climate change: Working group III contribution to the fifth assessment report of the Intergovernmental Panel on Climate Change*. New York, NY: Cambridge University Press.

IPCC, Intergovernmental Panel on Climate Change, 2014b. *Climate change 2014 synthesis report*. Geneva: IPCC.

ITDP, Institute for Transportation and Development Policy, 2013. *The bike-share planning guide*. New York, NY: ITPD.

Jackson, K.T., 1985. *Crabgrass frontier: The suburbanization of the United States*. Oxford: Oxford University Press.

Jacobson, S.H. and King, D.M., 2009. Fuel saving and ridesharing in the US: Motivations, limitations, and opportunities, *Transportation Research Part D*, Vol. 14 (1): 14–21.

Johnston, C., 2014. Uber's value more than doubles to $40bn after investors back fundraising, *The Guardian*, 5 December.

Jones, P., 2011. Conceptualising car 'dependence'. In: K. Lucas, E. Blumenburg and R. Weinberger, eds., *Auto motives: Understanding car use behaviours*. Bingley, UK: Emerald, 39–61.

Jussiant, L., 2002. Combined mobility and car-sharing, *Public Transport International*, Vol. 51 (6): 12–15.

Karlaftis, M.G., 2008. Privatisation, regulation, and competition: A thirty-year retrospective on transit efficiency. In: OECD and ITF, *Privatisation and regulation of urban transit systems*. Paris: OCED, 67–108.

Katzev, R., 2003. Car-sharing: A new approach to urban transportation problems, *Analyses of Social Issues and Public Policy*, Vol. 3 (1): 65–86.

Kay, J.H., 1997. *Asphalt nation: How the automobile took over America and how we can take it back*. Berkeley, CA: University of California Press.

Kenworthy, J.R., 2006. The eco-city: Ten key transport and planning dimensions for sustainable city development, *Environment and Urbanization*, Vol. 18 (1): 67–85.

Kerr, R., 2011. Peak oil production may already be here, *Science*, Vol. 331: 1510–1511.

Kopp, A., Block, R.I. and Iimi, A., 2013. *Turning the right corner: Ensuring development through a low-carbon transport sector*. Washington, DC: World Bank.

Kostakis, V. and Bauwens, M., 2014. *Network society and future scenarios for a collaborative economy*. Basingstoke, UK: Palgrave Macmillan.

Kuhnimhof, T., Zumkeller, D. and Chlond, B., 2013. Who made peak car, and how? A breakdown of trends over four decades in four countries, *Transport Reviews*, Vol. 33 (3): 325–342.

Kumar, A., Foster, V. and Barrett, F., 2008. *Stuck in traffic: Urban transport in Africa*. Washington, DC: World Bank.

Künneke, R. and Finger, M., 2009. The governance of infrastructures as common pool resources. *Fourth Workshop on the Workshop*, Bloomington, IN, June 2–7.

Kunstler, J.H., 2005. *The long emergency: Surviving the end of oil, climate change, and other catastrophes of the twenty-first century*. New York, NY: Grove Press.

Laws, R., Enoch, M., Ison, S. and Potter, S., 2009. Demand responsive transport: A review of schemes in England and Wales, *Journal of Public Transportation*, Vol. 12 (1): 19–37.

Leibowitz, S.J. and Margolois, S.E., 2002. The fable of the keys, *Journal of Law and Economics*, Vol. 30 (1): 1–26.

Leismann, K., Schmitt, M., Rohn, R. and Baedeker, C., 2013. Collaborative consumption: Towards a resource-saving consumption culture, *Resources,* Vol. 2 (3): 184–203.

Levosky, A. and Greenburg, A., 2001. *Organized dynamic ride sharing: The potential environmental benefits and the opportunity for advancing the concept*. Paper No. 01-0577. Transportation Research Board 2001 Annual Meeting, Washington, DC, January 7–11.

Lifkin, K., 2012. A whole new way of life: Ecovillages and revitalization of deep community. In: R. deYoung and T. Princen, eds., *The localization reader: Adapting to the coming downshift*. Cambridge, MA: MIT Press, 129–140.

Lifkin, K.T., 2014. *Ecovillages: Lessons for sustainable community*. Cambridge, UK: Polity Press.

Light, A., ed., 1998. *Social ecology after Bookchin*. New York, NY: Guildford Press.

Litman, T., 2005. *Well measured: Developing indicators for comprehensive and sustainable transport planning*. Victoria Transport Planning Institute.

Lovins, A.B., 2011. *Reinventing fire: Bold business solutions for the new energy era*. White River Junction, VT: Chelsea Green Publishing.

Low, N., ed., 2012. *Transforming urban transport: The ethics, politics and practices of sustainable mobility*. London: Routledge.

Luke, T., 1997. *Ecocritique: Contesting the politics of nature, economy, and culture*. Minneapolis, MN: University of Minnesota Press.

Luthra, A., 2006. Para transit system in medium sized cities: Problem or panacea, *ITPI Journal*, Vol. 3 (2): 55–61.

Macário, R., Viegas, J. and Hensher, D.A., eds., 2007. *Competition and ownership in land passenger transport*. Oxford: Elsevier.

Martell, L., 1994. *Ecology and society: An introduction*. Cambridge, UK: Polity Press.

Martin, E. and Shaheen, S., 2010a. *The impact of carsharing on household vehicle ownership*. University of California Transportation Center.

Martin, E. and Shaheen, S., 2010b. *Greenhouse gas emissions impacts of carsharing in North America*. MTI report 09-11. San José, CA: Mineta Transportation Institute.

Mattson, J., 2014. *Rural transit fact book 2014*. Tampa, FL: National Centre for Transit Research.

McDonough, M. and Braungart, W., 2009. *Cradle to cradle: Remaking the way we make things*. London: Vintage Books.

Meadows, D.H., Meadows, D.L., Randers, J. and Behrens, W.W., 1972. *The limits to growth*. New York, NY: Universe Books.

Meakin, R., 2004. *Urban transport institutions*. Sustainable Urban Transport Sourcebook, Module 1b. Eschborn: GTZ.

Mees, P., 2010. *Transport for suburbia: Beyond the automobile age*. London: Earthscan.
Menon, M., 2014. Uber driver in India arrested for allegedly raping female passenger, *Huffington Post*, 7 December 2014.
Metz, D., 2008. *The limits to travel: How far will you go?* London: Earthscan.
Metz, D., 2013. Peak car and beyond: The fourth era of travel, *Transport Reviews*, Vol. 33 (3): 255–270.
Metz, D., 2014. *Peak car: The future of travel*. London: Landor.
Midgley, P., 2011. *Bicycle-sharing schemes: Enhancing sustainable mobility in urban areas*. Background Paper No. 8. Commission on Sustainable Development, Nineteenth Session, New York.
Millard-Ball, A., Murray, G., Schure, J., Fox, C. and Burkhardt J., 2005. *Car-sharing: Where and how it succeeds*. Report 108. Washington, DC: Transportation Research Board.
Minister of Transport, 1963. *Traffic in towns*. London: HMSO.
Mirowski, P., 2013. *Never let a serious crisis go to waste: How neoliberalism survived the financial meltdown*. New York, NY: Verso.
Mobility Cooperative, 2015. *Business and sustainability report: Abridged version*. Lucerne: Mobility Cooperative.
Moriarty, P. and Honnery, D., 2008. The prospects for global green car mobility, *Journal of Cleaner Production*, Vol. 16 (16): 1717–1726.
Moriarty, P. and Honnery, D., 2010. *Rise and fall of the carbon civilisation: Resolving global environmental and resource problems*. London: Springer.
Mulley, C. and Nelson J.D., 2009. Flexible transport services: A new market opportunity for public transport, *Research in Transportation Economics*, Vol. 25 (1): 39–45.
Mumford, L., 1961. *The city in history: Its origins, its transformations, and its prospects*. New York, NY: Harcourt, Brace and World.
Mumford, L., 1967. *Technics and human development*. New York, NY: Harcourt Brace Jovanovich.
Murphy, P., 2013. *Plan C: Community survival strategies for peak oil and climate change*. Gabriola Island, BC: New Society Publishers.
Nader, R., 1965. *Unsafe at any speed: The designed-in dangers of the American automobile*. New York, NY: Grossman Publishers.
Newman, P. and Kenworthy, K., 2000. The ten myths of automobile dependence, *World Transport Policy and Practice*, Vol. 6 (1): 15–25.
Newman, P., Beatley, T. and Boyer, H., 2009. *Resilient cities: Responding to peak oil and climate change*. Washington, DC: Island Press.
Newman, P.W.G. and Kenworthy, J., 1999. *Sustainability and cities: Overcoming automobile dependence*. Washington, DC: Island Press.
Norcliffe, G., 2001. *The ride to modernity: The bicycle in Canada, 1869–1900*. Toronto: University of Toronto Press.
North, D., 1990. *Institutions, institutional change, and economic performance*. Cambridge, UK: Cambridge University Press.
Noya, A. and Lecamp, G., 1999. *Social enterprises*. Paris: OECD.
NRC, National Research Council, 2013. *Transitions to alternative vehicles and fuels*. Washington, DC: National Academies Press.
Nutley, S., 1996. Rural transport problems and non-car populations in the USA: A UK perspective, *Journal of Transport Geography*, Vol. 4 (2): 93–106.
Nutley, S., 2003. Indicators of transport and accessibility problems in rural Australia, *Journal of Transport Geography*, Vol. 11 (1): 55–71.

Nye, D.E., 1997. *Electrifying America: Social meanings of a new technology*. Cambridge, MA: MIT Press.

Nyssens, M., ed., 2006. *Social enterprise: At the crossroads of market, public policies and civil society*. Abingdon, Oxon: Routledge.

O'Connell, S., 2014. Motoring and mobility: 1900–1950. In: F. Carnevali and J.M. Strange, eds. *20th century Britain: Economic, cultural and social change*. London: Routledge, 111–126.

OECD, Organization for Economic Cooperation and Development, 1979. *Urban transport and environment*. 4 Vols. Paris: OECD.

OECD, Organisation for Economic Co-Operation and Development, 1999. *Social enterprises*. Paris: OECD.

OECD, Organisation for Economic Co-Operation and Development, 2002. *Policy instruments for achieving environmentally sustainable transport*. Paris: OECD.

OECD, Organisation for Economic Co-Operation and Development, 2003. *Privatising state-owned enterprises: An overview of policies and practices in OECD countries*. Paris: OECD.

OECD, Organisation for Economic Co-Operation and Development, 2008. *OECD Environmental outlook to 2030*. Paris: OECD.

OECD, Organisation for Economic Co-Operation and Development, 2014. *The cost of air pollution: Health impacts of road transport*. Paris: OECD.

OECD, Organisation for Economic Co-Operation and Development, 2015. *The metropolitan century: Understanding urbanization and its consequences*. Paris: OECD.

OECD/ITF, Organisation for Economic Co-Operation and Development/International Transport Forum, 2008. *Privatisation and regulation in urban transport systems*. Paris: OECD.

OECD/ITF, Organisation for Economic Co-Operation and Development/International Transport Forum, 2010. *Transport outlook 2010: The potential for innovation*. Paris: OECD.

OECD/ITF, Organisation for Economic Co-Operation and Development/International Transport Forum, 2013. *Cycling, health and safety*. Paris: OECD.

O'Riordan, T., 1981. *Environmentalism*. London: Pion.

Ostrom, E., 1990. *Governing the commons: The evolution of institutions for collective action*. Cambridge, UK: Cambridge University Press.

Owens, S., 1995. From 'predict and provide' to 'predict and prevent'? Pricing and planning in transport policy, *Transport Policy*, Vol. 2 (1): 43–49.

Paterson, M., 2007. *Automobile politics: Ecology and cultural political economy*. Cambridge, UK: Cambridge University Press.

Pearse, G., 2012. *Greenwash: Big brands and carbon scams*. Collingwood, Australia: Black Inc.

Pepper, D., 1993. *Eco-socialism: From deep ecology to social justice*. London: Routledge.

Pepper, D., 1996. *Modern environmentalism: An introduction*. London: Routledge.

Petersen, T., 2009. Network planning, Swiss-style: Making public transport work in semi-rural areas, *Proceedings of the 32nd Australasian Transport Research Forum*. Available at: www.patrec.org/web_docs/atrf/papers/2009/1802_paper72-Petersen.pdf

Petersen, T. and Mees, P., 2010. A case of good practice: The Swiss 'network' approach to semi-rural public transport, *12th World Conference on Transport Research*, Lisbon, Portugal, July 11–15.

Pfeiffer, D.A., 2006. *Eating fossil fuels: Oil, food and the coming crisis in agriculture*. Gabriola Island, BC: New Society Publishers.

Pierson, P., 2004. *Politics in time: History, institutions, and social analysis*. Princeton, NJ: Princeton University Press.

Ponting, C., 1991. *A green history of the world: The environment and the collapse of great civilizations*. Harmondsworth, UK: Penguin Books.

Posada, F., Kamakate, F. and Bandivadekar, A., 2011. *Sustainable management of two- and three-wheelers in Asia*. Working paper 2011-13. Washington, DC: The International Council on Clean Transportation.

Post, R.C., 2007. *Urban mass transit: The life story of a technology*. Westport, CT: Greenwood Press.

Prettenhaler, F.E. and Steininger, K.W., 1999. From ownership to service use lifestyle: The potential of car sharing, *Ecological Economics*, Vol. 28 (3): 443–453.

Pucher, J. and Buehler, R., eds., 2012. *City cycling*. Cambridge, MA: MIT Press.

Putnam, R., 2001. *Bowling alone: The collapse and revival of American community*. New York, NY: Simon and Schuster Paperbacks.

Rappoport, E., 2014. Utopian visions and real estate dreams: The eco-city past, present and future, *Geography Compass*, Vol. 8 (2): 137–149.

Register, R., 1987. *Ecocity Berkeley: Building cities for a healthy future*. Berkeley, CA: North Atlantic Books.

Reilly, M., 2014. Al Frankin confronts Uber after 'troubling' reports, *Huffington Post*, 19 November.

Ridley-Duff, R.J. and Bull, M., 2011. *Understanding social enterprise: Theory and practice*. London: Sage.

Roberts, P., 2005. *The end of oil: The decline of the petroleum economy and the rise of a new energy order*. London: Bloomsbury.

Robinson, J., 2004. Squaring the circle? Some thoughts on the idea of sustainable development, *Ecological Economics*, Vol. 48: 369–384.

Rogowsky, M., 2014. Are investors 'nuts' to value Uber at $18 billion? In a few years, that'll seem like a bargain, *Forbes*, 9 June.

Rome, A., 2001. *Bulldozer in the countryside: Suburban sprawl and the rise of American environmentalism*. Cambridge, UK: Cambridge University Press.

Romm, J.J., 2004. *The hype about hydrogen: Fact and fiction in the race to save the climate*. Washington, DC: Island Press.

Rooney, R.C., Bayley, S.E. and Schindler, D.W., 2012. Oil sands mining and reclamation cause massive loss of peatland and stored carbon, *Proceedings of the National academy of Sciences of the United States of America*, Vol. 109 (13): 4933–4937.

Rose, C., 1986. The comedy of the commons: Custom, commerce, and inherently public property, *University of Chicago Law Review*, Vol. 5 (3): 711–781.

Rose, G., 2012. E-bikes and urban transportation: Emerging issues and unresolved questions. *Transportation*, Vol. 39 (1): 81–96.

Roseland, M., 2012. *Toward sustainable communities: Solutions for citizens and their governments*. Fourth Edition. Gabriola Island, BC: New Society Publishers.

Ross, K., 1996. *Fast cars, clean bodies: Decolonization and the reordering of French culture*. Cambridge, MA: The MIT Press.

Ryden, C. and Morin, E., 2005. *Mobility services for urban sustainability: Environmental assessment*. Moses Report WP6, Trivector Traffic AB, Stockholm, Sweden.

Sachs, W., 1984. *For love of the automobile: Looking back into the history of our desires*. Berkeley, CA: University of California Press.

Safdie, M., 1998. *The city after the automobile*. Boulder, CO: Westview Press.

Sale, K., 1985. *Dwellers in the land: The bioregional vision*. San Francisco, CA: Sierra Club Books.

Salinger, L.W., ed., 2005. *Encyclopedia of white-collar and corporate crime*. Vol. 1. Thousand Oaks, CA: SAGE Publications.

Satterthwaite, D., 2007. *The transition to a predominantly urban world and its underpinnings*. IIED: London.

Schatzberg, E., 2001. Culture and technology in the city: Opposition to mechanized street transportation in late-nineteenth-century America. In: M.T. Allen and G. Hecht, eds., *Technologies of power: Essays in honor of Thomas Parke Hughes and Agatha Chipley Hughes*. Cambridge, MA: MIT Press, 57–94.

Schiller, P.L., Brun, E.C. and Kenworthy, J.R., 2010. *An introduction to sustainable transport: Policy, planning and implementation*. London: Earthscan.

Schipper, F., 2008. *Driving Europe: Building Europe on roads in the twentieth century*. Amsterdam: Aksant Academic Publishers.

Schipper, L., 2011. Automobile use, fuel economy and CO2 emissions in industrialized countries: Encouraging trends through 2008? *Transport Policy*, Vol. 18 (2): 358–372.

Schivelbusch, W., 1986. *The railway journey: The industrialization of time and space in the 19th century*. Berkeley, CA: The University of California Press.

Schneiderman, E.T., 2014. *Airbnb in the city*. New York, NY: New York State Office of the Attorney General.

Schumacher, E.F., 1973. *Small is beautiful: A study of economics as if people mattered*. London: Blond and Briggs.

Schwartz, B., 2004. *The paradox of choice: Why more is less*. New York, NY: Harper Perennial.

Seyfang, G., 2005. *Community currencies and social inclusion: A critical evaluation*. CSERGE Working Paper EDM 05-09, University of East Anglia, Norwich, UK.

Seyfang, G., 2009. *The new economics of sustainable consumption: Seeds of change*. Basingstoke: Palgrave Macmillan.

Seiler, C., 2009. *Republic of drivers: A cultural history of automobility*. Chicago, IL: University of Chicago Press.

Shaheen, S., Sperling, D. and Wagner, C., 1999. A short history of carsharing in the 90s, *The World Journal of Transport Policy and Practice*, Vol. 5 (3): 18–40.

Shaheen, S., Guzman, S. and Zhang, H., 2010. *Bikesharing in Europe, the Americas, and Asia: Past, present, and future*. Washington, DC: Transportation Research Board Annual Meeting.

Shaheen, S.A. and Cohen, A.P., 2013. Carsharing and personal vehicle services: Worldwide market developments and emerging trends, *International Journal of Sustainable Transportation*, Vol. 7 (1): 5–34.

Shaheen, S. and Christensen, M., 2014. *Shared use mobility summit: Retrospectives from North America's first gathering on shared-use mobility*. Transportation Sustainability Research Center, University of California Berkeley.

Shaheen, S.A., Zhang, H., Martin, M. and Guzman, S., 2011. China's Hangzhou public bicycle: Understanding early adoption and behavioral response to bikesharing, *Transportation Research Record*, No. 2247: 33–41.

Shaheen, S.A., Elliot, W.M., Chan, N.D., Cohen, A.P. and Pogodzinski, M., 2014. *Public bikesharing in North America during a period of rapid expansion: Understanding business models, industry trends and user impacts*. MTI report 12-29. San José, CA: Mineta Transportation Institute.

Sharma, B.R., 2008. Road traffic injuries: A major global public health crisis, *Public Health*, Vol. 122 (12): 1399–1406.

Sheller, M. and Urry, J., 2006. The new mobilities paradigm, *Environment and Planning A*, Vol. 38 (2): 207–226.

Siegal, L., 2008. *Against the machine: Being human in the age of the electronic mob*. New York, NY: Spiegal and Grau.

Simpson, B.J., 1996. Deregulation and privatization: The British local bus industry following the Transport Act 1985, *Transport Reviews*, Vol. 16 (3): 213–223.

SLoCaT (Partnership on Sustainable Low Carbon Transport), 2013. *Creating universal access to safe, clean and affordable transport: A status report on the contribution of sustainable transport to the implementation of Rio+20*. Available at: http://slocat.net/sites/default/files/u10/slocat_status_report_rio_20-_june_19_2013_1.pdf

Spielvogel, J.J., 2012. *Western civilization since 1300*. Eighth Edition. Boston, MA: Wadsworth.

Social Exclusion Unit, 2003. *Making the connections: Final report on transport and social exclusion*. London: Social Exclusion Unit.

Soppelsa, P.S., 2009. *The fragility of modernity: Infrastructure and everyday life in Paris, 1870–1914*, PhD Dissertation, The University of Michigan, ProQuest, UMI Publishing Service.

Soron, D., 2009. Driven to drive: Cars and the problem of 'compulsory consumption'. In J. Conley and A.T. McLaren, eds., *Car troubles: Critical studies of automobility and auto-mobility*. Farnham: Ashgate, 181–196.

Sperling, D. and Gordon, D., 2009. *Two billion cars: Driving towards sustainability*. Oxford: Oxford University Press.

Squires, G.D., 2002. *Urban sprawl: Causes, consequences and policy responses*. Washington, DC: The Urban Institute Press.

Stough, R.R. and Reitveld, P., 1997. Institutional issues in transport systems, *Journal of Transport Geography*, Vol. 5 (3): 207–214.

Summers, L.H. and Balls, E., 2015. *Report of the Commission on Inclusive Prosperity*. Washington, DC: Centre for American Progress.

Taylor, N., 2000. Eco-villages: Dream and reality. In: H. Barton, ed., *Sustainable communities: The potential for eco-neighbourhoods*. Abingdon, OX: Earthscan, 19–28.

Tilman, D., Socolow, R., Foley, J.A., Hill, J., Larson, E., Lynd, L., Pacala, S., Reilly, J., Searchinger, T., Somerville, C. and Williams, R., 2009. Beneficial biofuels: The food, energy, and environment trilemma, *Science*, Vol. 325: 270–271.

Trainer, T., 1995. *The conserver society*. London: Zed Books.

Trieb, O., Bähr, H. and Falkner, G., 2007. Modes of governance: Towards a conceptual clarification, *Journal of European Public Policy*, Vol. 14 (1): 1–20.

Tumlin, J., 2011. *Sustainable transport planning: Tools for creating vibrant, healthy, and resilient communities*. Hoboken, NJ: Wiley.

UN, United Nations, 2008. *World urbanization prospects: The 2007 revision*. New York, NY: United Nations.

UN, United Nations, 2014. *The world population situation in 2014: A concise report*. New York, NY: United Nations.

UN-Habitat, 2013. *Global report on human settlements 2013: Planning and design for sustainable urban mobility*. Abingdon, UK: Ashgate.

UNDESA, United Nations, Dept. of Economic and Social Affairs, 2014. *World urbanization prospects: The 2014 revision: Highlights*. New York, NY: United Nations.

UNEP, United Nations Environment Programme, 2014a. *The emissions gap report 2014*. Nairobi: UNEP.

UNEP, United Nations Environment Programme, 2014b. *Yearbook: Emerging Issues in our Global Environment 2014*. Nairobi: UNEP.

UNFCCC, United Nations Framework Convention on Climate Change, 2009. *Report on the conference of the parties on its fifteenth session, held in Copenhagen from December 7–19, 2009*. Available at: http://unfccc.int/resource/docs/2009/cop15/eng/11a01.pdf

UNPF, United Nations Population Fund, 2007. *State of the world population: Unleashing the potential of urban growth*. New York, NY: UNPF.

USEIA, United States Energy Information Administration, 2013. *International energy outlook 2013: With projections to 2040*. Washington, DC: US Department of Energy.

Urry, J., 2007. *Mobilities*. Cambridge, UK: Polity Press.

Urry, J., 2008. Governance, flows, and the end of the car system? *Global Environmental Change*, Vol. 18 (3): 343–349.

Van Dender, K. and Clever, M., 2013. *Trends in car usage in advanced countries: Slower growth ahead: Summary and conclusions*. Discussion paper No. 2013–9. Paris: ITF/OECD.

Van de Velde, D.M., 1999. Organizational forms and entrepreneurship in public transport, *Transport Policy*, Vol. 6 (3): 147–157.

Vasconcellos, E.A., 2013. *Transport, environment and equity: The case for developing countries*. Abingdon, OX: Earthscan.

Velaga, N.R., Nelson, J.D., Wright, S.D. and Farrington, J.D., 2012. The potential of flexible transport services in enhancing rural public transport provision, *Journal of Public Transportation*, Vol. 15 (1): 111–131.

Vuchic, V.R., 2007. *Urban transit systems and technology*. Hoboken, NJ: John Wiley and Sons.

Walker, J., 2012. *Human transit: How clearer thinking about public transit can enrich our communities and our lives*. Washington, DC: Island Press.

Wall, D., 2014. *The commons in history: Culture, conflict, and ecology*. Cambridge, MA: MIT Press.

WBCSD, World Business Council for Sustainable Development, 2004. *Mobility 2030: Meeting the challenges to sustainability: Executive summary*. Geneva: WBCSD.

WBCSD, World Business Council for Sustainable Development, 2009. *Mobility for development*. Geneva: WBCSD.

WCED, World Commission on Environment and Development, 1987. *Our common future*. Oxford: Oxford University Press.

Weber M., 2001. *The protestant ethic and the spirit of capitalism*. Trans. S. Kalberg, Chicago, IL: Roxbury.

Weinert, J., Ma, C. and Cherry, C., 2007. The transition to electric bikes in China: History and key reasons for rapid growth, *Transportation*, Vol. 34 (3): 301–318.

Weinert, J., Ogden, J., Sperling, D. and Burke, A., 2008. The future of electric two-wheelers and electric vehicles in China, *Energy Policy*, Vol. 36 (7): 2544–2555.

WHO, World Health Organization, 2013a. *Global status report on road safety 2013: Supporting a decade of action*. Geneva: WHO.

WHO, World Health Organization, 2013b. IARC: Outdoor air pollution a leading environmental cause of cancer deaths. *Press Release no. 221*, 17 October.

von Weizsächer, E., Lovins, A.B. and Lovins, L.H., 1998. *Factor four: Doubling wealth, halving resource use*. London: Earthscan.

Whitelegg, J., 1993. Time pollution, *Ecologist*, Vol. 23 (4): 131–134.

Wills-Johnson, N., 2010. Railway dreaming: Lessons for economic regulators from Aboriginal resource management lore, *Review of Policy Research*, Vol. 27 (1): 47–58.

Wolfe, J., 2010. *Autos and progress: The Brazilian search for modernity*. Oxford, UK: Oxford University Press.

Woods. M., 2005. *Rural geography: Processes, responses and experiences in rural restructuring*. London: Sage.

World Bank, 2005. *A study of institutional, financial and regulatory frameworks of urban transport in large sub-Saharan cities*. SSATP Working Paper No. 82. Washington, DC: World Bank.

World Bank, 2010. *World development indicators 2010: Development and climate change*. Washington, DC: World Bank.

World Economic Forum, 2012. *Sustainable transport ecosystem: Addressing sustainability from an integrated systems perspective*. Geneva: World Economic Forum.

Worldwatch Institute, 2012. *Biofuels for transport: Global potential and implications for sustainable energy and agriculture*. London: Earthscan.

Wright, C. and Curtis, B., 2005. Reshaping the motor car, *Transport Policy*, Vol. 12 (1): 11–22.

Yago, G., 1984. *The decline of transit: Urban transportation in German and U.S. cities: 1900–1970*. Cambridge, UK: Cambridge University Press.

Index

For Product Safety Concerns and Information please contact our EU
representative GPSR@taylorandfrancis.com
Taylor & Francis Verlag GmbH, Kaufingerstraße 24, 80331 München, Germany

www.ingramcontent.com/pod-product-compliance
Ingram Content Group UK Ltd.
Pitfield, Milton Keynes, MK11 3LW, UK
UKHW021009180425
457613UK00019B/866

* 9 7 8 0 3 6 7 1 3 8 8 3 7 *